SpringerBriefs in Molecular Science

Chemistry of Foods

Series Editor

Salvatore Parisi, Lourdes Matha Institute of Hotel Management and Catering
Technology, Thiruvananthapuram, Kerala, India

The series Springer Briefs in Molecular Science: Chemistry of Foods presents compact topical volumes in the area of food chemistry. The series has a clear focus on the chemistry and chemical aspects of foods, topics such as the physics or biology of foods are not part of its scope. The Briefs volumes in the series aim at presenting chemical background information or an introduction and clear-cut overview on the chemistry related to specific topics in this area. Typical topics thus include:

- Compound classes in foods—their chemistry and properties with respect to the foods (e.g. sugars, proteins, fats, minerals, …)
- Contaminants and additives in foods—their chemistry and chemical transformations
- Chemical analysis and monitoring of foods
- Chemical transformations in foods, evolution and alterations of chemicals in foods, interactions between food and its packaging materials, chemical aspects of the food production processes
- Chemistry and the food industry—from safety protocols to modern food production

The treated subjects will particularly appeal to professionals and researchers concerned with food chemistry. Many volume topics address professionals and current problems in the food industry, but will also be interesting for readers generally concerned with the chemistry of foods. With the unique format and character of SpringerBriefs (50 to 125 pages), the volumes are compact and easily digestible. Briefs allow authors to present their ideas and readers to absorb them with minimal time investment. Briefs will be published as part of Springer's eBook collection, with millions of users worldwide. In addition, Briefs will be available for individual print and electronic purchase. Briefs are characterized by fast, global electronic dissemination, standard publishing contracts, easy-to-use manuscript preparation and formatting guidelines, and expedited production schedules.

Both solicited and unsolicited manuscripts focusing on food chemistry are considered for publication in this series. Submitted manuscripts will be reviewed and decided by the series editor, Prof. Dr. Salvatore Parisi.

To submit a proposal or request further information, please contact Dr. Sofia Costa, Publishing Editor, via sofia.costa@springer.com or Prof. Dr. Salvatore Parisi, Book Series Editor, via drparisi@inwind.it or drsalparisi5@gmail.com

Subhabrata Panda

Soil and Water Conservation for Sustainable Food Production

 Springer

Subhabrata Panda
All India Coordinated Research Project on
Agroforestry, Department of Soil and Water
Conservation
Bidhan Chandra Krishi Viswavidyalaya
West Bengal, India

ISSN 2191-5407   ISSN 2191-5415   (electronic)
SpringerBriefs in Molecular Science
ISSN 2199-689X   ISSN 2199-7209   (electronic)
Chemistry of Foods
ISBN 978-3-031-15404-1   ISBN 978-3-031-15405-8   (eBook)
https://doi.org/10.1007/978-3-031-15405-8

This Springer imprint is published by the registered company Springer Nature Switzerland AG
The registered company address is: Gewerbestrasse 11, 6330 Cham, Switzerland

Series Editor's Foreword

This book is intended to be a constituent companion of structured textbooks on soil science, soil conservation, soil physics, soil health, soil and water conservation, etc. A single book cannot afford to deal with all aspects of knowledge of soil science, as that branch of science has grown up to the unthinkable dimensions from studies on massive soil mass to soil particles and colloidal clay micelle, soil microbial community, soil organic matter, soil nutrients, soil moisture and their uptake by plants, management of aquatic environment. Moreover, physical, chemical, microbial and biochemical properties of soils and their interactions with climate and hydrological conditions for successful crop cultivation should be considered in this ambit. Naturally, all the relevant books have to deal with the application of methodologies for determining all those vast soil features, from gravimetric methods to applications of nuclear and nanotechnologies, and application of soil and water conservation methods. Basically, the correct management of soil organic matter and soil moisture would diminish soil loss and simplify the management of problem soils and irrigation water with the aim of reaching the 15 Sustainable Development Goals of the United Nations. Consequently, this book presents a brief discourse on the development of basic ideas concerning above-mentioned areas of concepts and applications. Hopefully, this book will be a ready reference for students appearing in competitive exams and internship evaluation projects and serve as a brief commentary for studies, research and field works targeted on individual agricultural plots and further development of soil and water conservation as a full-fledged stream of science.

Palermo, Italy

Salvatore Parisi
Series Editor for SpringerBriefs
in Chemistry of Foods

Acknowledgements

I am really happy to express my deepest sense of gratitude to Prof. Salvatore Parisi for offering me an opportunity to write this book, and his continuous reminder for completion of this work has encouraged me to recover from an infection of COVID-19, a global pandemic, and after damage of my computer system due to a lightning incident.

I am also equally grateful to all the staff associated with the publishing of books from the Springer Nature, especially Charlotte Hollingworth, Sofia Costa, Christoph Baumann, Thomas Hempfling, Selma Somogy, Stephanie Kolb, Vidyaa Shri Krishna Kumar, Antje Endemann, Cansu Kaya, Ravi Vengadachalam, Monica Janet M and others associated with the Springer Science Business Science + Media Deutschland GmbH for creating a friendly atmosphere with continuous reminder to the author for completion of writing the manuscript.

I also feel indebted to all my teachers: Prof. D. K. Datta, Dr. A. K. Ghosh, Prof. R. K. Ghose, Late Prof. P. De, Prof. A. K. Chakravarti, Prof. R. K. Biswas and all the staff of the erstwhile Department of Agricultural Engineering, BCKV; my seniors Prof. P. K. Dhara, Prof. N. C. Das, Prof. R. Ray of the present Department of Soil and Water Conservation, BCKV, for their continuous encouragement during my studies and research works in the erstwhile Department of Agricultural Engineering; with sincere thanks to my colleague Prof. S. K. De for his fellow feelings. I hereby extend my sincere regards to all scientists, staff, especially to Joydev Rana, Prdip Kumar Nayek, Dr. B. Biswas with the AICRP on Agroforestry and Regional Research Station (Red & Laterite Zone), BCKV, Jhargram, West Bengal, India.

I also feel thankful to Dada (Ardhendu Sekhar Mishra), Boudi (Kalpana Mishra), Smritirekha, Abir, Aditya, Bappaditya, Piu, Binapani, Bodhi, Deep, Jhuma, Yashita and all my family members for extending their help for creating a nice atmosphere in my home.

I am appreciative to Pinakesh Das, Dr. Udita Mondal Mukherjee (Assistant Professor, Brainware University), Anirban Bhowmik, Moumita Khatun, Rajesh Pradhan, Subha Mollah, Sambhunath Saren, Biswajit Saren, Milan Hembram and other research and past postgraduate students for their inquisitiveness on problems and applications of field and laboratory studies on soil and water conservation technologies as an encouragement to write this book.

I hereby convey my special thanks to Deepankar Dutta of ITECHNOsavvy, Subhankar Roy of Connectica Management Service, Dr. Bappaditya Mishra of Velarudh Infotech Pvt. Ltd. for creation of a hassle-free computer system for this work, Deepankar Dutta and Abir Panda for computer drawing of few diagrams and Abir Panda and Smritirekha Panda for sharing some photographs from their field works.

Jhargram, India   Subhabrata Panda
June 2022

Contents

1 Introductory Remarks: Soil and Water Conservation for Soil
  Health 1
  1.1 Introduction 2
  1.2 Summary 7
  References 8

2 Soil Properties Responsible for Soil Loss 13
  2.1 Introduction 15
  2.2 Erodibility of Soil and Its Three-Phase System 17
    2.2.1 Erodibility of Soil and Soil Texture 17
    2.2.2 Erodibility of Soil and Soil Structure 17
  2.3 Soil Properties Responsible for Water Erosion 18
    2.3.1 Soil Properties and Estimation of Water Erosion 18
  2.4 Soil Properties Responsible for Both Water and Wind Erosions 20
    2.4.1 Forces of Wind/Water Acting on Soil Grains
          at Threshold of Soil Movement 21
  2.5 Soil Properties Influencing Crop Growth 22
  2.6 Soil Loss Causing Vulnerability to Soil Ecosystem—A Major
      Dialectics of Nature 26
  2.7 Concluding Remarks 29
  References 30

3 Impact of Climate, Water and Biological Factors on Soil Health 35
  3.1 Introduction 36
  3.2 Formulation of Soil Quality Indexing Including Water
      Quality in the Context of Climate Change 38
  3.3 Effect of Water on Soil Health in the Context of Climate
      Change 39
  3.4 Soil Microbial Biomass—A Tool for Assessment of Soil
      Health in the Context of Climate Change 43
  3.5 Genetic and Functional Biodiversity of Soils, Soil Health
      and Climate Change 44

  3.6   Soil Health Key Indicators for in Situ Soil Health Assessment
        Under Climate Change 45
  3.7   Concluding Remarks 46
  References 46

4 **Effect of Soil on Water Quality** 53
  4.1   Introduction 54
  4.2   Effect of Geology on Water Quality 54
  4.3   Effect of Topography on Water Quality 55
  4.4   Effect of Soil Erosion and Water Quality as Influenced
        by Climate 55
  4.5   Effect of Soil Properties on Water Quality 55
  4.6   Effect of Soil Erosion on Water Quality as Influenced
        by Vegetation Cover 56
  4.7   Effect of Watershed on Water Quality 56
        4.7.1   Effect of Soil Erosion on Water Quality of Aquatic
                Ecosystem and Watershed Hydrology 57
  4.8   Effect of Land Use Land Cover on Water Quality 60
  4.9   Concluding Remarks 61
  References 61

5 **Soil and Water Qualities Necessary for Irrigation** 65
  5.1   Introduction 66
  5.2   Land Characterisation Necessary for Irrigation 66
  5.3   Soil and Water Compatibility Necessary for Irrigation 67
        5.3.1   Interaction Between Soil and Water 67
        5.3.2   Physiological Drought Soil Condition 69
        5.3.3   Diagnosis of Soil Properties for Irrigation Management 71
        5.3.4   Irrigation Water Quality 73
  5.4   Irrigation Management in Salt Affected Soils 74
  5.5   Diagnosis of Salt Affected Soils 75
  5.6   Soil and Water Management for Sustainable Crop Production 76
  5.7   Concluding Remarks 76
  References 77

6 **Soil Moisture Conservation Influencing Food Production** 79
  6.1   Introduction 80
  6.2   Soil Moisture Storage as Affected by Rooting Depth, Soil
        Bulk Density, Rainfall and Evapotranspiration 81
  6.3   Soil Moisture Conservation Techniques and Implementation 82
        6.3.1   Implementation Considerations for Soil Moisture
                Conservation Technologies 86
  6.4   Beneficial Roles of Soil Moisture Conservation 88
  6.5   Prospects and Problems of Soil Moisture Conservation
        Techniques 88
        6.5.1   Prospects of Soil Moisture Conservation Techniques 88

      6.5.2   Problems of Soil Moisture Conservation Techniques   89
  6.6   Concluding Remarks   89
  References   89

7   Management of Soil Organic Carbon   91
  7.1   Introduction   92
  7.2   SOM, Carbon, Nitrogen, Phosphorus, Sulphur and Humus
       Interrelations   94
  7.3   Role of Soil Organic Matter on Soil Aggregate Stability   96
  7.4   Concluding Remarks   100
  References   100

8   Concluding Remarks: Soil and Water for Food Security   103
  8.1   Concluding Remarks   103
  References   106

About the Author

Subhabrata Panda is Assistant Professor in Soil and Water Conservation at the Bidhan Chandra Krishi Viswavidyalaya (BCKV), the State Agricultural University in West Bengal, India, and he is associated with the "All India Coordinated Research Project on Agroforestry" of the Indian Council of Agricultural Research (ICAR)-Central Agroforestry Research Institute (CAFRI). He has previously worked as Assistant Agricultural Chemist at the State Agricultural Research Institute and additionally with the Soil Survey Laboratory and State Nodal Cell of the Soil Health Management (SHM) under National Mission for Sustainable Agriculture (NMSA), Government of West Bengal, Kolkata, India, and acted as Research Associate in the ICAR-MNES All India Net Work Project on Solar Photovoltaic Pumping Systems for crop water management and in the ICAR-sponsored inter-institutional Arsenic Project with BCKV on studying mobilisation and management of arsenic from groundwater in Nadia district in West Bengal, India.

Chapter 1
Introductory Remarks: Soil and Water Conservation for Soil Health

Abstract Better soil health is the manifestation of the perpetuation of eight soil functions by continuously overcoming ten soil threats with single goal of achieving sustainable food production in future through adoption of rational policy based on application of the soil and water conservation science. Awareness generation is to be created among policy-makers and soil and water governance authorities as well not to accept sorrowful big projects again and for usefulness for application of site-specific scientific logic applicable for individual crop fields or farm for implementing small-scale measures of soil and water conservation as an independent fully fledged branch of science.

Keywords Agriculture · Climate · Food production · Soil health · Soil organic matter · Sustainable development goal · Water

Abbreviations

CEC	Commission of the European Communities
FAO	Food and Agriculture Organization of the United Nations
GIS	Geographic information system
ITPS	Intergovernmental Technical Panel on Soils
MEA	Millennium Ecosystem Assessment
SOM	Soil Organic Matter
SDG	Sustainable Development Goal
SSM	Sustainable soil management
UN	United Nations
UNCCD	United Nations Convention to Combat Desertification
WWC	World Water Council

1.1   Introduction

Soil and water conservation and soil health are now becoming indistinguishable in the present context of growing incidents of environmental pollutions as well as their ill effects on human health. Traditionally, soil is exclusively considered as the seat of food production as well as dumping spaces of all throwaways. Such conventional misuses of soil are needed to be regulated in the present situation of multiple environmental crises, mostly of anthropogenic origins. Such critical present period in environmental conditions requires necessary attention towards achieving Sustainable Development Goals (SDG) for securing Zero Hunger (SDG 2) as well as access to safe water and sanitation (SDG 6) for all (UN 2021). Those two criteria are essentially required for better human health and for attaining those goals for which soil is considered as one of the important factors for affording food security (SDG 2) through food production as well as making provisions for more water in rural areas (SDG 6) through rainwater harvesting. In this way, all our efforts towards food security are related to both soil and water; and their healthy uses.

Misuse or abuse of land resources also affects soil and water qualities which, in turn, affects our health. Food crops grown on contaminated soil or irrigated through contaminated water give contaminated food through built-up of pollutants in the crop produce. Soil and water resources are necessary either for crop production or for affording safe drinking water and water needed for other health purposes, and there may be multiple causes for pollution of those valuable assets. Application of various measures of soil and water conservation is needed for judicious use of those two inseparable natural resources as well as their quantitative and qualitative conservation for controlling natural or man-made losses for which maintaining soil health is the prerequisite factor. In this way, soil and water conservation and soil health have become synonymous in the present day of environmental crises.

Nothing has become successful and even the changed environment in altering our conventional behaviour of exclusively considering soil in the context of food production, though during past two decades new innovative ways have emerged on relating soil with social issues like food security and with other environmental issues like greenhouse gas emission and soil erosion and connecting soil with overall human well-being as recognised in the Status of the World's Soil Resources (FAO and ITPS 2015).

Like flesh and blood, we are obviously used to consider soil and water as the unified entity in the natural system. So, functions of soil in the ecosystem are the joint manifestations of both soil and water in that system. Seven soil functions, so far, recognised are (1) biomass production, including agriculture and forestry, (2) storing, filtering and transforming nutrients, substances and water, (3) biodiversity pool, such as habitats, species and genes, (4) physical and cultural environment for humans and human activities, (5) source of raw materials, (6) acting as a carbon pool and (7) archive of geological and archaeological heritage. The proposed Soil Framework Directive (CEC 2006) of the European Union has admitted those soil functions as vulnerable to soil threats mostly due to anthropogenic interferences in the ecosystem in present times. Consequently, with regard to follow the scientific basis for conservation and sustainable use of ecosystem, the UN Millennium Ecosystem Assessment (MEA 2005) is applicable for assessing consequences of ecosystem change for human well-being, based on the framework for ecosystem services (Costanza et al. 1997; Daily et al. 1997). That ecosystem assessment is categorised into four broad classes of (I) supporting, (II) regulating, (III) provisioning and (IV) cultural services. Considering soils as the main media of food production, the major ecosystem services of soil are mainly: (I) supporting services with the specific soil functions for enabling two services: (1) primary production with soil functions of (i) medium for seed germination and root growth, and (ii) supply of nutrients and water for crops; (2) nutrient cycling with soil functions of (i) transformations of organic matter by soil organisms and (ii) retention and release of nutrients on ionised soil particle surface; (II) The regulating services of soil in the ecosystem with four specific soil functions (1) water quality regulation with soil functions of (i) filtering and buffering of substances in soil water and (ii) transformation of contaminants; (2) water supply regulation with soil functions of (i) regulation of water infiltration and permeability into soil and within the soil, and (ii) drainage of excess water out of soil either as groundwater or as surface runoff; (3) climate regulation with soil function of regulation of emissions of carbon dioxide, nitrous oxide and methane gases from soil; and (4) erosion regulation with soil function of retention of soil on the land surface; (III) The provisioning services of soil related to food production with two soil functions are as follows: (1) food supply—providing water, nutrients and physical support for growth of plants for human and animal consumption, and (2) water supply—retention and purification of water. Apart from those, soil formation ecosystem services are related to soil genesis and are supported by four soil functions: (i) weathering of primary minerals and release of nutrients, (ii) transformation and accumulation of organic matter, (iii) creation of soil structures (aggregates, horizons) for flow of water and gases and root growth, and (iv) creation of charged soil particle surface for ion retention and exchange (Fig. 1.1).

Bockheim and coworkers have classified the developments in the complexity of soil knowledge from pre-1880 to 2015 into seven chronological periods with the evolutions of different soil management practices with soil conservation initiated in the period of 1900–1940 followed by sustainable soil management practices based on scientific assessments of soil functions in the following times (Bockheim et al. 2005). After 2015, it is found that agricultural systems are now required to be continually adapted and optimised through changes in cropping and land use management in the present day of environmental concerns of general public, climate change, droughts and floods (Ahuja et al. 2019). To cope up with the situation, quantitative whole system approach is required to be followed through application of computer-based advance science and technology, e.g., remote sensing, geographic information systems (GIS), artificial intelligence, etc. In this regard, the present age may be termed as digital period with biophysical system modelling (Fig. 1.2).

In the present-day context of climate change, small-scale system management will remain as the only answer based on the Nature-based Solution for managing soil and water system as a whole. On updating all previous understandings, including a recent

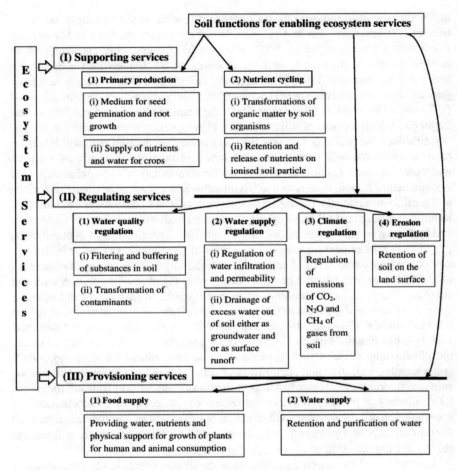

Fig. 1.1 Ecosystem services and soil functions (based on MEA 2005)

Communication from the Commission to the Council, the European Parliament, the European Economic and Social Committee and the Committee of the Regions (CEC 2006) and on considering the effect of each trouble on particular soil functions, the Status of the World's Soil Resources (FAO and ITPS 2015) has identified ten threats to soil: (1) soil acidification, (2) soil salinization, (3) soil biodiversity loss, (4) nutrient imbalance, (5) soil erosion, (6) soil organic carbon loss, (7) soil contamination, (8) soil sealing, (9) soil compaction and (10) waterlogging. Those soil threats come in the way of good soil health with recognised eight soil functions for plant productions, viz. (i) transformation and accumulation of soil organic matter (SOM), (ii) supply of nutrients, (iii) creation of charged surfaces, (iv) transformation of contaminants, (v) medium of seed generation and plant growth, (vi) regulation of water infiltration, (vii) retention and supply of water and (viii) drainage of excess water (Fig. 1.3).

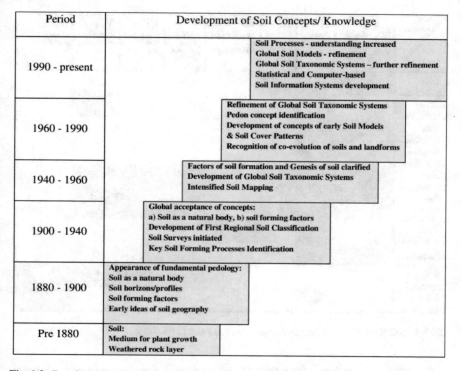

Fig. 1.2 Developments in soil knowledge from pre-1880 to present times (based on Bockheim et al. 2005)

Soil erosion still remains to be one of the ten major soil threats as influenced by the major drivers of soil organic carbon loss, soil biodiversity loss and soil degradation. It is anticipated that as a consequence of anthropogenic activities accelerated rates of soil erosion will continue in the coming decades. On the other hand, it is also well thought of that soil conservation practices and sustainable soil management (SSM) can reduce the rates of soil erosion in critically human-induced erosion-prone areas (Borrelli et al. 2020; FAO 2019a; Gharibreza et al. 2020; Guerra et al. 2020; Lefèvre et al. 2020; Naipal et al. 2018; Olson et al. 2016; Tsymbarovich et al. 2020).

Identification and solving a problem depend on execution of a scientifically sound policy framework and governance of the situation. This is true in execution of soil and water conservation measures for controlling soil erosion, soil degradation and saving of both the soil and water resources including their qualities towards sustainable soil health, as outlined in national (van Leeuwen et al. 2019; Xie et al. 2019), regional (Turpin et al. 2017) and international agreements and programmes (FAO 2017); Voluntary Guidelines for Sustainable Soil Management or UNCCD Land Degradation Neutrality concept (UNCCD 2017; Orr et al. 2017). Achieving at least half of the SDG is predicted by reducing and restoring eroded soils (FAO 2019b).

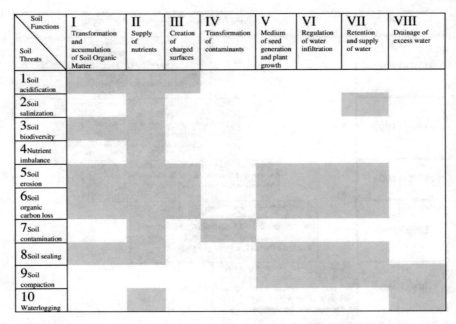

Fig. 1.3   Soil threats on soil functions (based on FAO and ITPS 2015)

Therefore, exercising soil and water conservation measures requires application of improved science with accurate estimates of soil erosion losses at increased spatial and temporal resolutions, and policy execution based on better understanding of science of soil and water conservation and data sharing between partners with common goal (FAO 2019b) for achieving sound soil health for sustainable food production. It was found that small-scale farm level irrigation schemes like farm ditches and percolation tanks should be followed on priority basis because those are more economic (Om-kar et al. 1975) and useful and potential enough within small watersheds. Those are suitable for farm water management in Pakistan (Bergsma 1985), important in India (Caprihan 1975) and useful for water conservation in arid and semi-arid areas (Limaye 1975). National Commission on Agriculture (1976) in India had proposed for maximum conservation of rainwater in ponds in West Bengal, India, where farm ponds were found suitable for transforming a monocropped area into a yearlong cultivable tract for agricultural production in three crop seasons (Panda 1987; Panda and De 1989; Panda et al. 1990; Saren 2019). From a water quality study, it was proved that water from farm ponds was more suitable for irrigation than any other sources in salt-affected soils in the Sundarban areas in West Bengal, India (Dhara 1986; Dhara et al. 1991). Consequently, encouraging policy effects of water harvesting in India were also revealed (Ray 1986). Those storages of collected rainfall and runoff in rural areas were also described as a great hope for thousands of scattered, small communities in the Third World that could not be served by more centralised water supply schemes in the foreseeable future (Pacey and

Cullis 1986), and those were found suitable for village potable water in Guatemala (Salazar 1983). In countries like India, detention of flashy rainfall of short duration by constructing battery of such mini-storages would be an ideal scheme to iron out the nonuniform spread of rainfall and also for recharging the underground sub-soil by infiltration. Those farm ponds were found effective on flood retardation and provision for supplemental irrigation from harvested water in various parts of India (Narayana 1993). It was proved that traditional decentralised methods of water management were better than centralised modern management due to hydrological characteristics of the concerned areas (Slaheddine 1986). Those research activities were initiated in the wake of sorrowful experiences for flood control, irrigation and food production as well as devastating losses of land, life and livelihoods resulting from big projects constructed in the last century (Connell 2013; Ghosh 2014; Luino et al. 2014). Those projects have failed in conserving soil and water for which countries are now again facing recurrent floods and droughts year after year due to uncontrollable heavy sedimentation. But, in recent times, the issue of all sorts of infrastructures is again raised with a pushing towards technology (i.e., capital)-intensive smarter agriculture in the business model with the allowance of private sector and creation of scopes for incentivising of such hopeless efforts with foreign finances (Hirji and Davis 2009) though as a whole small-scale approaches could not be overlooked (FAO 2011). So, in the back drop of past experiences, awareness is required to be generated among policy-makers and authorities for governance of soil and water resources based on scientific logic and usefulness for implementing small-scale measures of soil and water conservation for soil health towards sustainable food production to support a global population of 9 to 10 billion (FAO 2021; FAO and WWC 2015; NREL 2021).

1.2 Summary

From the detailed discourse, it is found accepted that better soil health is the manifestation of the perpetuation of eight soil functions by continuously overcoming ten soil threats with single goal of achieving sustainable food production in future through adoption of rational policy based on application of the soil and water conservation science. It is so argued because soil water conservation measures must include both mechanical, biological as well as ameliorative and crop cultural measures for protecting soil from erosion and qualitative degradation due to soil compaction, soil acidity, salinity, waterlogging, etc. Though water and wind are the primary agents of soil erosion, anthropogenic causes like deforestation, intensive cultivation, misuse of irrigation water, mismanagement of cultivated lands, overgrazing and urbanisation are also some of the potential drivers of soil degradation. In this regard, soil and water conservation is the application of science to be understood for selecting site-specific suitable measures among either mechanical, cultural including biological and ameliorative measures for correcting the soil physicochemical properties or combination of those as per necessity and social acceptability. Awareness generation is to be created among policy-makers and soil and water governance authorities not to accept sorrowful big projects again and for usefulness for application of site-specific scientific logic for implementing small-scale measures of soil and water conservation as an independent fully fledged branch of science for ensuring soil health, water conservation, flood retardation, saving agricultural lands from degradation, provision of potable water in rural areas and irrigation for sustainable food production. So, emphasis is given on discourses in every chapter about studies on soil and water problems and prospects and conservation of both these natural resources in relation to individual crop fields or farms.

References

Ahuja LR, Ma L, Anapalli SS (2019) Biophysical system models advance agricultural research and technology: some examples and further research needs, In: Wendroth O, Lascano RJ, Ma L (eds) Bridging among disciplines by synthesizing soil and plant processes, Chap 1, within Ahuja LR (Series Ed) Advances in agricultural systems modeling, vol 8. ASA, CSSA, SSSA, Madison. https://doi.org/10.2134/advagricsystmodel8.2017.0008

Bergsma HM (1985) Training and the development of curriculum standards in on farm water management: Pakistan, 1984–1985. Department of Curriculum and Instruction, New Mexico State University, Las Cruces, and Department of Curriculum and Instruction, Colorado State University, Ft. Collins

Bockheim JG, Gennadiyev AN, Hammer RD, Tandarich JP (2005) Historical development of key concepts in pedology. Geoderma 124(1–2):23–36. https://doi.org/10.1016/j.geoderma.2004.03.004

Borrelli P, Robinson DA, Panagos P, Lugato E, Yang JE, Alewell C, Wuepper D, Montanarella L, Ballabio C (2020) Land use and climate change impacts on global soil erosion by water (2015–2070). Proc NatlAcad Sci 117(36):21994–22001. https://doi.org/10.1073/pnas.2001403117

Caprihan SP (1975) Past, present and future of irrigation practices in Madhya Pradesh in India. Water for human needs, Proceeding 2nd world congress on water resources, Indian committee of IWRA, CBIP I, New Delhi, pp 253–263

CEC (2006) Communication from the commission to the council, the European parliament, the European economic and social committee and the committee of the regions. Thematic strategy for soil protection. COM 231 final. Commission of the European communities (CEC), Brussels. Available https://eur-lex.europa.eu/legal-content/EN/TXT/PDF/?uri=CELEX:52006D C0231&from=EN. Accessed 16 June 2022

Connel D (2013) The Tennessee valley authority: catchment planning for social development. Part of an 11-part series titled 'International water politics'. Global water forum (https://globalwaterf orum.org/). Available https://globalwaterforum.org/2013/03/20/international-water-politics-the-tennessee-valley-authority-catchment-planning-for-social-development/. Accessed 16 June 2022

Costanza R, d'Arge R, de Groot R, Farber S, Grasso M, Hannon B, Limburg K, NaeemS ORV, Paruelo J, Raskin RG, Sutton P, van den Belt M (1997) The value of the world's ecosystem services and natural capital. Nature 387(6630):253–260. https://doi.org/10.1038/387253a0

Daily GC, Matson PA, Vitousek PM (1997) Ecosystem services supplied by soil. In: Daily G (ed) Nature's services: societal dependence on natural ecosystems. Island Press, Washington, DC, pp 113–132

Dhara PK, Panda S, Roy GB, Datta DK (1991) Effect of waterbodies on the quality of groundwater in coastal areas of South 24 Parganas in West Bengal, India. J Ind Soc Coastal Agric Res 9(1/2):395–396

Dhara PK (1986) Study on the quality of water available from different sources in Sundarban area. Dissertation, West Bengal, India

FAO and ITPS (2015) Status of world's soil resources (SWSR)—main report. Intergovernmental technical panel on soils (ITPS), and FAO, Rome. Available https://www.fao.org/policy-support/tools-and-publications/resources-details/en/c/435200/. Accessed 16 June 2022

FAO and WWC (2015) Towards a water and food secure future—critical perspective of policy-makers. Food and agriculture organization of the United Nations, Rome, and world water council (WWC), Marseille. Available http://www.fao.org/3/i4560e/i4560e.pdf. Accessed 16 June 2022

FAO (2011) The state of the world's land and water resources for food and agriculture (SOLAW)—managing systems at risk. Food and agriculture organization of the United Nations (FAO), Rome, and Earthscan, London. Available http://www.fao.org/nr/solaw/the-book/en/. Accessed 16 June 2022

FAO (2017) Voluntary guidelines for sustainable soil management. Food and agriculture organization of the United Nations, Rome. Available https://www.fao.org/3/bl813e/bl813e.pdf. Accessed 16 June 2022

FAO (2019a) Soil erosion: the greatest challenge for sustainable soil management. Food and agriculture organization of the United Nations, Rome. Available https://www.fao.org/3/ca4395en/ca4395en.pdf. Accessed 16 June 2022

FAO (2019b). Outcome document of the global symposium on soil erosion. Food and agriculture organization of the United Nations, Rome. Available https://www.fao.org/3/ca5697en/ca5697en.pdf. Accessed 16 June 2022

FAO (2021) Land and water: sustainable land management. Food and agriculture organization of the United Nations, Rome. Available

Gharibreza M, Zaman M, Porto P, Fulajtar E, Parsaei L, Eisaei H (2020) Assessment of deforestation impact on soil erosion in loess formation using 137Cs method (case study: Golestan Province, Iran). Int J Soil Water Conserv Res 8:393–405. https://doi.org/10.1016/j.iswcr.2020.07.006

Ghosh S (2014) The impact of the Damodar valley project on the environmental sustainability of the lower Damodar basin in West Bengal, Eastern India. OIDA Int J Sustain Develop 7(2):47–54. Available http://www.ssrn.com/link/OIDA-Intl-Journal-Sustainable-Dev.html. Accessed 18 June 2022

Guerra CA, Rosa IMD, Valentini E, Wolf F, Filipponi F, Karger DN, Nguyen Xuan A, Mathieu J, Lavelle P, Eisenhauer N (2020) Global vulnerability of soil ecosystems to erosion. Landsc Ecol 35(4):823–842. https://doi.org/10.1007/s10980-020-00984-z

Hirji R, Davis R (2009) Environmental flows in water resources policies, plans, and projects: findings and recommendations. Environment and development. World Bank, Washington, DC. Available https://openknowledge.worldbank.org/handle/10986/2635. Accessed 18 June 2022

https://www.fao.org/land-water/land/sustainable-land-management/en/. Accessed 16 June 2022

Lefèvre C, Cruse RM, Cunha dos Anjos LH, Calzolari Haregeweyn CN (2020) Guest editorial—soil erosion assessment, tools and data: a special issue from the Global Symposium on soil Erosion 2019. Int Soil Water Conserv Res 8(4):333–336. https://doi.org/10.1016/j.iswcr.2020.11.004

Limaye SD (1975) Utility of percolation tanks for water conservation in arid and semi-arid regions. In: Water for human needs, Proceedings of the second world congress on water resources, Indian committee of IWRA. CBIP IV, New Delhi, pp 205–208

Luino F, Tosatti G and Bonaria V (2014) Dam failures in the 20th century: nearly 1000 avoidable victims in Italy alone. J Environ Sci Eng A 3(1):19–31

MEA (2005) Ecosystems and human well-being: synthesis. Millennium ecosystem assessment (MEA), World Resources Institute, Washington, DC. Island Press, Washington, DC

Naipal V, Ciais P, Wang Y, Lauerwald R, Guenet B, Oost KV (2018) Global soil organic carbon removal by water erosion under climate change and land use change during AD 1850 e 2005. Biogeosci 15:4459e4480. https://doi.org/10.5194/bg-15-4459-2018

Narayana VVD (1993) Water harvesting, recycling and estimation of runoff. In: Narayana VVD (ed) Soil and water conservation research in India, Chap. 8. ICAR, New Delhi, pp 111–151

National Commission on Agriculture (1976) Rainfall and cropping patterns, vol XVI. Ministry of Agriculture and Irrigation, West Bengal, Govt. of India, New Delhi

NREL (2021) Food security. Natural resources ecology lab (NREL), Fort Collins, Colorado. Available https://www.nrel.colostate.edu/research/food-security/. Accessed 18 June 2022

Olson KR, Al-Kaisi M, Lal R, Cihacek L (2016) Impact of soil erosion on soil organic carbon stocks. J Soil Water Conserv 71(3):61A-67A. https://doi.org/10.2489/jswc.71.3.61A

Om-kar et al (1975) Production costs in irrigated agriculture in the lower Mekong Basin. In: Water for human needs, Proceedings of the second world congress on water resources, Indian Committee of IWRA. CBIP I, New Delhi, pp 265–278

Orr BJ,Cowie AL,Castillo Sanchez VM,Chasek P,Crossman ND, Erlewein A, Louwagie G,Maron M, Metternicht GI, Minelli S, Tengberg AE, Walter S, Welton S (2017) Scientific conceptual framework for land degradation neutrality.A report of the science-policy interface.United Nations Convention to Combat Desertification (UNCCD), Bonn

Pacey A, Cullis A (1986) Rainwater harvesting: the collection of rainfall and runoff in rural areas. Intermediate Technology Publications, London

Panda S, Roy GB, Ghosh RK (1990) Detection of agroclimatic feasibility for transforming an apparently water deficit monocropped area to a yearlong cultivable tract in Contai, Midnapore, West Bengal. Indian J Landsc Sys Ecol Stud 13(2):174–175. https://doi.org/10.13140/RG.2.2. 14316.41600

Panda S, De P (1989) Hydrological balance approach for planning irrigation in coastal areas of Contai, Midnapore, West Bengal. In: Proceedings of all India seminar on "Role of hydrology in efficient management of irrigation system", organised by Indian association of hydrologists, West Bengal regional centre. At the Auditorium, Birla Industrial and Technological Museum, 19 A, Calcutta 700019, West Bengal, India, 19–20 May (1989)

Panda S (1987) Water management planning: a case study in a coastal mauza (Contai-III, Dist-Midnapore, West Bengal). Dissertation, M.Sc. (Agriculture) in soil and water conservation thesis, Department of Agricultural Engineering, BCKV, West Bengal, India

Ray D (1986) Some agricultural policy effects of encouraging water harvesting in India. Agric Adm 21(4):235–248. https://doi.org/10.1016/0309-586X(86)90041-5

Salazar LJ (1983) Water management on small farm: a training manual for farmers in hill area. Water management synthesis project no. 88, University Services Centre, Colorado State University, Fort Collins

Saren S (2019) Mouza wise water management planning in red and laterite zone of West Bengal. Dissertation, M.Sc. (Agriculture) in soil and water conservation thesis, Bidhan Chandra Krishi Viswavidyalaya, West Bengal, India

Slaheddine EA (1986) Traditional versus modern irrigation methods in Tunisia. In: Goldsmith E, Hildyard N (eds) The social and environmental effects of large dams, vol 2. Case studies. Wadebridge Ecological Centre, Camel-Ford Cornwall

Tsymbarovich P, Kust G, Kumani M, Golosov V, Andreeva O (2020) Soil erosion: an important indicator for the assessment of land degradation neutrality in Russia. Int Soil Water Conserv Res 8(4):418–429. https://doi.org/10.1016/j.iswcr.2020.06.002

Turpin N, ten Berge H, Grignani C, Guzman G, Vanderlinden K, Steinmann HH, Siebielec G, Spiegel A, Perret E, Ruysschaert G, Laguna A, Gir aldez JV, Werner M, Raschke I, Zavattaro L, Costamagna C, Schlatter N, Berthold H, Sanden T, Baumgarten A (2017) An assessment of policies affecting sustainable soil management in Europe and selected member states. Land Use Policy 66:241e249. https://doi.org/10.1016/j.landusepol.2017.04.001

UN (2021) The sustainable development goals report. The sustainable development goals website of the United Nations, Department of Economic and Social Affairs Statistics Division, The United Nations, New York. Available https://unstats.un.org/sdgs/report/2021/. Accessed 18 June 2022

UNCCD (2017) Land degradation neutrality transformative action—tapping opportunities. Global mechanism of the UNCCD Platz der Vereinten Nationen, Bonn. Available https://www.unccd. int/sites/default/files/documents/2017-10/171006_LDN_TP_web.pdf. Accessed 18 June 2022

van Leeuwen CC, Cammeraat EL, de Vente J, Boix-Fayos C (2019) The evolution of soil conservation policies targeting land abandonment and soil erosion in Spain: a review. Land Use Policy 83:174–186. https://doi.org/10.1016/j.landusepol.2019.01.018

Xie Y, Lin H, Ye Y, Ren X (2019) Changes in soil erosion in cropland in northeastern China over the past 300 years. Catena 176:410e418. https://doi.org/10.1016/j.catena.2019.01.026

Chapter 2
Soil Properties Responsible for Soil Loss

Abstract Soil structure influences retention and flow of water, solutes, gases and distribution of biota in soil. Soil aggregate or soil structure stability is the determining factor of soil loss, an unavoidable natural phenomenon occurring on all landforms. Agents of soil particle cohesiveness, adhesiveness with water and soil organic matter and other biological factors control soil loss. Anthropogenic factors like soil cultural practices and other misuses of soil enhance loss of soil. Recognising soil erosiveness as function of some soil properties would be most helpful in identifying various lands with variations in soil erodibility under various land use practices for minimising soil loss from individual crop fields, and this will be helpful in achieving Sustainable Development Goals towards food security.

Keywords Dispersion ratio · Soil erodibility · Soil loss · Soil organic carbon · Soil organic matter · Sustainable development goal · Water

Abbreviations

AD	Aggregate density
Agg.	Aggregation index
Al_2O_3	Aluminium oxide
Ag	Amount of > 200 mm-aggregate
$(A + LF)_{Max}$	Maximum amount of dispersed 0–20 mm fraction obtained after three treatments of the initial soil sample
ASD	Dry aggregate size distribution
CEC	Cation exchange capacity
CDR	High clay dispersion ratio
C_i	Final concentration of the stabilising constituent (organo-mineral complexes or humic fraction)
C_o	Original concentration of the stabilising constituent (organo-mineral complexes or humic fraction)
Cl	Percent of clay
DAS	Dry aggregate stability

© The Author(s), under exclusive license to Springer Nature Switzerland AG 2022
S. Panda, *Soil and Water Conservation for Sustainable Food Production*,
Chemistry of Foods, https://doi.org/10.1007/978-3-031-15405-8_2

DR	Dispersion ratio
D_r	Range of particle size
EI	Erosion Index
FAO	Food and Agriculture Organization of the United Nations
Fe_2O_3	Iron (III) oxide
Fed	Dithionite extractible iron
GIS	Geographical information system
I_C	Index of crusting
I_e	Index of erodibility
I_r	Index of resistance
I_s	Instability index
I_{ss}	Index of structural Stability
K	Soil erodibility
k	Soil permeability
k_1	Rate constant
LDN	Land degradation neutrality
NPP	Net primary productivity
R_2O_3	Sesquioxide
RUSLE	Revised USLE
S_c	Percent of coarse silt
S_f	Percent of fine silt
S	Percent of sand
si	Percent of silt
S_t	Critical soil organic matter content
SG	Content of coarse mineral sands (>200 μm)
SiO_2	Silica
SWAT	Soil and Water Assessment Tool
SE	Soil erosion
SOC	Soil organic carbon
SOM	Soil organic matter
SDG	Sustainable Development Goal
T_i	Time
UNCCD	United Nations Convention to Combat Desertification
UNESCO	United Nations Educational, Scientific and Cultural Organization
UN	United Nations
USDA	United States Department of Agriculture
USLE	Universal Soil Loss Equation
W	Soil moisture content
WDC	Water dispersible clay
ρ_B	Soil bulk density

2.1 Introduction

Soil loss occurs from all landforms. It starts from loosening of soil mass within soil matrix itself by the actions of erosive agents like water and wind as influenced by climatic, hydrological, biological and anthropogenic factors, as those are operative in weathering of rocks towards formation of soils (Fig. 2.1). Consolidated soil mass, i.e., soil matrix is formed as soil particles are closely packed through various physical and biological agents. Soil matrix originates from binding of soil particles by cohesive forces and also through adhesive forces acting between soil minerals and nonsoil mineral materials. Colloidal soil particles are bound among themselves by electrical properties originating from within colloidal particles. In that way, very fine soil particles of colloidal size, i.e., the size of the dispersed phase of a colloidal system, are bound together; and as a result, soil colloidal particles may help to bring other soil granules of bigger sizes to some extent and other nonsoil mineral materials into a soil matrix. Also, attractions in between soil–water, soil–organic matter, etc., act as binding agents within soil mass as adhesive forces. Among those, moisture and biological exudates mostly help as gum to bind soils and other materials within the soil matrix. Other factors of climate (humidity), gravitation (e.g., weight of a soil matrix), hydrological (e.g., formation of ice sheet, infiltration of water through soil surface) and biological (e.g., decomposition and humification of organic matter through decaying of floral and faunal masses within soil, and addition of exudates from living roots and dead organic matter) are some of the examples of other agents which act for binding soil particles by adhesive forces. A combination of those forces is active for plant roots to physically bind with soil masses. Weakening of those cohesive and adhesive forces acting within soil mass is the cause of loosening of soil particles and ultimate loss of soil from the ground. In that way, physical forces of water, wind and anthropogenic factors are the main causes of soil loss. After considerable works on discovery of principles of erosiveness of soil, it has been revealed that by keeping all factors equal, some soils are more readily erodible than others. That has prompted to search for a fundamental principle being responsible for producing such a difference in soil loss among various types of soils. From a review on works of Middleton and his associates (Middleton 1930; Middleton et al. 1932, 1934) for their thorough studies on soil physical and chemical properties influencing soil erosion and based on those studies, it is revealed that 'Middleton has proposed erosion ratio as the best single criterion of erosion' (Bouyoucos 1935). Consequently, from recent works it is to be noted that soil structure influences retention and flow of water, solutes, gases and distribution of biota in soil; but the ubiquitous heterogeneity in soil structure cannot be described easily with the use of soil aggregates (Young et al. 2001), though it is not yet solved. On the other hand, there are some empirical models like USLE (Universal Soil Loss Equation), RUSLE (Revised USLE) and SWAT (Soil and Water Assessment Tool) to estimate soil loss from a geographical area, but availability and quality of data are major constraints against finding out soil loss from a place (Golosov et al. 2019). In that context, recognising soil erosiveness as functions of some soil properties would be of most helpful in identifying various lands with variations in soil erodibility under various land use practices for minimising soil loss from individual plots.

Fig. 2.1 Weathered rocks

2.2 Erodibility of Soil and Its Three-Phase System

Soil is a three-phase system of solid, liquid and gases interconnected through capillary pores that enable the movement of water in soil by capillary forces. Those pore spaces are filled with liquid or gas phases. Solid phase wets in contact with water and is covered with a thin film of it. Sometimes, organic matter may cover the dry soil particles for which soil can exert limited hydrophilicity or can be temporarily water repellent, though plant cannot grow in permanent porous media. Susceptibility of soil to erosion is characterised by the dynamic attribute of erodibility of soil. This is an estimate of Erosion Index (EI) as obtained through laboratory or plot or field studies. Erodibility of soil is affected by soil properties like texture, structure, organic matter content, hydraulic properties and wettability.

2.2.1 Erodibility of Soil and Soil Texture

Erodibility of soil is its susceptibility or vulnerability to erosion. It can be expressed as a function of both physical characteristics of soil and its management in a prevailing climate and landscape situation. Erodibility of soil is indirectly correlated with cohesiveness of soil aggregates in contrast with water infiltration through soil. Well-aggregated clayey soils are more cohesive than sandy soils, and thus, clayey soils are more resistant to erosion than sandy soils. Though due to smaller size, clay particles are easily removed by runoff after loosening from soil aggregates. Silty soils of loess parent material origin are most erodible types of soils. This is in contrast with the water infiltration because water infiltration increases with coarser soil aggregates, i.e., with less cohesive soil aggregates. Macropores can conduct water more rapidly than micropores. Coarse soils with low cohesive soil aggregates are with high water infiltration capacity, though sandy soils with more macropores have less combined pore spaces than clayey soils. Under low intensity rainfall clayey soils produce more runoff due to its more cohesive aggregates (Hudson 1981; Lal et al. 2006; Wuest et al. 2006).

2.2.2 Erodibility of Soil and Soil Structure

Soil structure is the mechanical framework of soil particles and biological substances including pore spaces within soil matrix. Stability of soil structure is indicated by air permeability, organic matter content, soil moisture content, water infiltration, etc. Soils with poor structures are easily erodible with low water infiltration capacity and susceptibility to compaction. Those macroscale structural attributes determine stability of soil for retaining water, storing organic matter and withstanding erosion. Following investigation by tomography and necessary explanation by fractal theory, soil fragmentation due to tillage operation, susceptibility of soil to erosion and water flow through soil are governed by the abundance, arrangement and complexity of soil particles and soil pores. (Young et al. 2001).

2.3 Soil Properties Responsible for Water Erosion

Raindrop splashes cause detachment of clay particles and, in turn, exposed soil surface is broken and, thereby, dispersed clay particles, being detached from running water, form a thin film of compact deposit of clay cover, i.e., surface sealing. That surface sealing reduces rate of infiltration of water into soil and increase in runoff rate as well as increase in transport of soil, i.e., increase in water erosion. The dispersed fine clay particles after settling down cause clogging of soil pores for which infiltration rate decreases which causes increase in runoff and soil erosion. Factors influencing surface sealing are soil aggregate stability, soil surface cover by vegetation and mulch, soil texture, soil organic matter content, incipient soil moisture status, rain drop impact and energy, and other anthropogenic factors including tillage management. Surface sealing on drying appears as soil crusts of about 0.1 to 5 cm thickness (Blanco-Canqui and Lal 2008; Gao et al. 2005; Lu et al. 2016; Ma et al. 2014).

Soil aggregates, described by Baver as 'soil microstructure' or secondary particle (Baver 1935), are formed from (1) the finer mechanical separates and (2) amount and effectiveness of the granulating or aggregating substances present within the soil. Then, granulating substances are identified as organic matter, iron and aluminium compounds and other similar compounds. Baver has found significant correlation between aggregation and clay percentage, organic matter and carbon (Baver 1935). Soil organic matter (SOM) is one of the major causes of soil granulation, excepting lateritic soil where colloidal aluminium oxide (Al_2O_3) and iron (III) oxide (Fe_2O_3) significantly affect soil aggregate formation. In this way, soil aggregate stability can be a function of soil colloidal particles, SOM and cations present in soil. The European Union, based on an appraisal on 'Soil and water in a changing environment', has identified interrelations among different soil parameters and indicators of pedotransfer functions, as shown in Table 2.1 (European Union 2014).

2.3.1 Soil Properties and Estimation of Water Erosion

Most erodible soils are generally high in silt, low in clay and low in organic matter (Wischmeier and Mannering 1969). Regardless of corresponding increase in sand or clay fractions, soil types become less erodible with decrease in sand fraction. According to them, percentages of silt, clay and sand must be considered in relation to existing levels of other physical and chemical properties of soil, and they have denoted aggregation index as sand:silt ratio (Eq. 2.1).

Table 2.1 Soil parameters, key explanatory parameter used in soil water retention curve (SWRC) and indicator as estimated using PTFs, i.e., pedotransfer functions (based on European Union 2014)

Soil parameters	Key explanatory parameter	Indicator as estimated using pedotransfer functions
Texture	% Silt % Clay % Sand	Granulometric distribution
Compactness and structure	Bulk density (BD)	% OM, % clay, % silt, topsoil, subsoil
Cation exchange capacity	% Organic matter (SOM)	Organic carbon content (SOC), total N
	Hydraulic conductivity	Saturated hydraulic conductivity

$$\text{Aggregation index} = \frac{\text{Sand}}{\text{Silt}} \tag{2.1}$$

Wischmeier and Mannering (1969) proposed following two relations (Eqs. 2.2 and 2.3) combining soil erodibility (K), reaction (R), percent of sand (s), clay (c) and silt (si), aggregation index (agg.) and clay ratio:

$$K \propto \left(0.043R + \frac{0.62}{OM} + 0.0082s - 0.0062c \right) si \tag{2.2}$$

and

$$K \propto \left(1.73 - 0.26OM - \frac{2.42}{OM} + 0.3agg. - \frac{0.024}{agg.} - 0.0062si \right) \text{clay ratio} \tag{2.3}$$

Bouyoucos has proposed clay ratio as an index of erosiveness of soil, as shown in Eq. (2.4) (Bouyoucos 1935):

$$\text{Index of erosiveness or Clay ratio} = \frac{(\text{Sand} + \text{Silt})}{\text{Clay}} \tag{2.4}$$

The ratio of percentages of silt and clay determined by dispersion in distilled water to total silt and clay obtained (Eq. 2.5) through mechanical analysis is termed as dispersion ratio (Middleton 1930; Lal and Shukla 2004; Singh 1980):

$$\text{Dispersion ratio} = \frac{\text{Dispersed(Silt} + \text{Clay)}}{\text{Total(Silt} + \text{Clay)}} \tag{2.5}$$

Now, (Middleton 1930):

$$\text{Erosion ratio} = \text{Dispersion ratio} / \left(\frac{\text{Colloid content}}{\text{Moisture equivalent}} \right) \tag{2.6}$$

Here, moisture equivalent is the soil moisture content when soil is subjected to a centrifugal force equivalent to 1000 G (Briggs and McLane 1907; Briggs and Shantz 1912; Dakshinamurti and Gupta 1968; Veihmeyer and Hendrickson 1931).

It has been found that moisture equivalent, lower liquid limit, maximum water holding capacity, slaking value and shrinkage follow closely the mechanical composition and colloid contently of soil (Middleton 1930). For tropical soils, dispersion ratio (DR) is only moderately high and most often it may range as low as 0.13–0.94. Soils with high DR are more erodible and conversely less erodible soils are with low DR values. DR with a value of 10 is considered to be a boundary between erodible and nonerodible soils. Erodibility of soils is related to level of sesquioxides, organic matter and presence of other binding agents in soil mass (Glinski et al. 2011; Igwe 2005; Lal and Shukla 2004).

2.4   Soil Properties Responsible for Both Water and Wind Erosions

Soil erosion can be defined as the loosening of soil aggregates followed by 'the detachment and transport of soil particles by erosive agents, most commonly water and/or wind' (Flanagan et al. 2013). Generally, erosion of soil by wind/ water is the function of original soil property, detachment and transport properties of erosive agents in interactions with climate, landscape position and management practices (Cruse et al. 1990). Erodibility of soil greatly influences susceptibility of a land to erosion. Similar to water erosion, a number of empirical and physically based models is developed for prediction of wind erosion, but availability and quality data remain the main constraint. So, such obstacles can be removed through finding out soil properties responsible for wind erosion also, which will be helpful for specifying a landscape with certain extent of wind erosion required for adopting precautionary or ameliorative land management practices to minimise soil erosion. Soil erosion by water and wind is a serious problem throughout the world. About 75 billion tons of topsoil are lost every year from agricultural areas throughout the world (Flanagan et al. 2013; Myers 1993), whereas soils generally take thousands of years from their original parent material through the course of natural (i.e., geological or normal) erosion as a part of that soil development process (Flanagan et al. 2013). That normal soil formation rate is in the order of 25 mm in 300 to 1000 years (Bennett 1939; Pimental et al 1976). Aeration and leaching actions are speeded up by tilling of land for which soil formation rate becomes 25 mm in 30 years. So, a target figure of 11.21 t ha^{-1} year^{-1} soil erosion should not exceed (Hudson 1981) though recently reported sustainable rate of soil loss (i.e., equal to the rate of soil formation) is about 1 t ha^{-1} year^{-1} (FAO 2000). It is also reported that 16% of total world land area (21,960,000 km^2 of 134,907,000 km^2) is subjected to significant soil erosion risk. Rate of soil loss is greatest in Asia, South America and Africa with average value of about 30–40 t ha^{-1} year^{-1} and that for North America and Europe at about 17 t ha^{-1} year^{-1}(FAO 2000; Flanagan et al. 2013).

To find out wind erodibility of soil, most important factors recognised are soil structure including size, shape and density of erodible and nonerodible fractions (Chepil and Woodruff 1963) as quantified through dry aggregate stability (DAS), dry aggregate size distribution (ASD), aggregate density (AD), incipient surface moisture content, crusting and loose erodible material on the crust surface (Zobeck et al. 2013).

Disintegration and detachment of soil aggregates into separate particles are essentially required preconditions for movement of soil particles by wind. That movement is initiated as the wind pressure overcomes gravitational force against loose surface soil grains, followed by movement of soil particles in a series of jumps known as saltation. Impact of saltating grains initiates movement of larger and denser grains and causes disintegration of massive materials or clods due to collision. Wind erosion of soil occurs only when soil grains capable of being moved by saltation comparatively few saltating grains jump higher than a few feet above the ground and 90% do not rise above 1 foot, and thus, wind erosion is a surface phenomenon extending to saltation height. The higher the soil particles jump they gain more energy with the increase in saltating grains concentration (i.e., number per unit volume) downwind as the eroding field is large enough and it becomes maximum until the velocity of wind can sustain to flow with increased load of grains and incipient air pressure. A wind strong enough to initiate is always turbulent, and soil abrasion occurs firstly due to saltation. 'Lowering of wind velocity due to soil movement varies directly with soil erodibility; i.e., more the erodible soil, the greater the concentration of moving soil grains and the greater is the reduction of wind velocity near the ground' (Chepil and Woodruff 1963).

2.4.1 Forces of Wind/Water Acting on Soil Grains at Threshold of Soil Movement

A moving fluid either wind/ water exerts three types of forces on soil particles resting on ground (Chepil and Woodruff 1963; Einstein and El-Samni 1949; Ippen and Verma 1953):

(1) Positive pressure against the part of grain facing into the direction of fluid motion. That pressure, here, is the force per unit of cross-sectional area of the grain normal to the direction of fluid motion, and it varies directly with the square of the fluid velocity. That pressure results from the impact of the fluid against the grain and is called the impact or velocity pressure for which movement of soil grains is initiated.

(2) Negative pressure on the lee side of the grain is known as viscosity pressure. Its magnitude is dependent on fluid's coefficient of viscosity, density and velocity.

Fig. 2.2 Wind erosion **a**: the Rajasthan desert and **b** the Bay of Bengal sea shore

(3) Negative pressure on the top, as compared to the bottom, of the grain. It is
    caused by Bernoulli effect and is called static, isotropic or internal pressure.
    According to Bernoulli law, wherever the fluid velocity is speeded up, as at
    the top of soil grain, the pressure (as measured to the general direction of fluid
    motion) is reduced (Chepil and Woodruff 1963).

Impact or velocity pressure on soil grain lying on the ground is the cause of the
formation of form drag. That drag and the pressure due to viscous shear in the fluid
close to the surface of the soil grain is unitedly called skin friction drag. The sum of
those two forces is the total drag, which acts on the top of the grain at the threshold of
its movement due to the pressure difference against its windward and leeward sides.
A decrease in static pressure at the top of the grain as compared to that at the bottom
causes a lift on the grain. It is determined by, but is not as the same as, the pressure
difference against the top and the bottom halves of the grain and it acts through the
centre of the gravity (Chepil and Woodruff 1963).

Regarding motion of sediment particles in water, nothing is confirmed about
turbulent pressure fluctuations near stream bed. But actions of raindrop splash, sheet,
rill and gully erosions might be due to turbulent pressure generated for such types of
water erosions. Such turbulent pressure might have link with stream bank erosions,
in contrast with movement of sediment grains on the bed surface.

So, initiation of soil erosion by wind or water is theoretically identical, only
excepting at the stream bed which can be moved by applying artificial dragging force,
for example by dredging machine, by which processes of water erosion ultimately
resembles wind erosion at the threshold of the movement of soil grains (Figs. 2.2,
2.3, 2.4 and 2.5).

2.5 Soil Properties Influencing Crop Growth

The orientation of soil particles in soil structure largely determines physical characteristics of soil as well as its efficacy in crop growth through the eventual pore size
distribution in soil matrix. Soil structure helps to continue qualitative and quantitative relation with solid, liquid and gas—i.e., three phases of soil system. Optimal

Fig. 2.3 Raindrop splash and sheet erosions

Fig. 2.4 Rill erosion

Fig. 2.5 Gully erosion

condition of such interrelation helps to maintain exchange with environment in a suitable reciprocal way to meet up the necessities of soil microorganisms and plant roots. Consequently, it requires mention that pore size distribution is important in specifying a soil structure and depends on bonding and arrangements of soil aggregates. Development of interparticle bonds on the other hand is helpful for stability of soil aggregates. In this context, the following are also required to be noted:

(1) Aggregated (silt + clay), an index of erodibility of soil (Middleton 1930). This is computed from the analysis done for dispersion ratio. It is the difference between actual (silt + clay) and the percent suspension without dispersion.

(2) Surface aggregation ratio (Anderson 1954) as given in Eq. (2.7):

$$\text{Surface aggregation ratio} = \frac{\text{Total surface area of particles} > 0.05 \text{ mm}}{\text{Aggregated (silt + clay)}}$$

(2.7)

(3) Index of resistance, I_r (Chorley 1959), as displayed in Eq. (2.8):

$$I_r = (\rho_b x \times D_r)/w \tag{2.8}$$

where ρ_b soil bulk density, D_r is the range of particle size and w is soil moisture content.

(4) Index of Erodibility, I_e (Chorley 1959), as displayed in Eq. (2.9):

$$I_e = (I_r \times k)^{-1} \tag{2.9}$$

where I_r is index of resistance and k is soil permeability.

(5) Index of Structural Stability, I_{ss} (Kay et al 1988), based on change in level of stabilising material, as displayed in Eq. (2.10):

$$I_{ss} = C_i/C_o = \left(1 - e^{-k_1 T_i}\right) \tag{2.10}$$

where C is the stabilising constituent (organo-mineral complexes or humic fraction) representing original (C_o) and final (C_i) concentrations, T_i is the time (year) and k_1 is the rate constant.

(6) Instability index, I_s, based on texture and cementing agents involved in aggregation of tropical soils, as displayed in Eq. (2.11) (Hénin et al. 1958):

$$I_s = \frac{(A + LF)_{max}}{\frac{1}{3}Ag - 0.9SG} \tag{2.11}$$

where $(A + LF)_{max}$ is the maximum amount of dispersed 0–20 mm fraction obtained after three treatments of the initial soil sample: (i) air dry soil (without any pretreatments), (ii) followed by immersion in alcohol and (iii) then immersion in benzene; Ag is the amount of >200 mm-aggregates (air, alcohol, benzene) obtained after shaking (30 manual turnings and wet sieving of those three pretreated soil samples); SG is content of coarse mineral sands (>200 μm); and (1/3 Ag − 0.9SG) is the mean stable aggregates.

(7) Index of Crusting, I_C, based on soil textural composition and soil organic matter content, as written in Eq. (2.12) (Lal and Shukla 2004):

$$I_C = \frac{1.5S_f + 0.75S_c}{Cl + (10 \times SOM)} \tag{2.12}$$

where S_f is percent fine silt, S_c is percent coarse silt, Cl is percent clay and SOM is percent soil organic matter content. It is obvious that I_C is inversely related to clay and soil organic matter, and directly to fine and coarse silt contents.

(8) Critical soil organic matter content, S_t, proposed by Pieri, based on the concept (Pieri 1991) of critical level of soil organic matter concentration for structural stability of tropical soils as the organic matter concentration plays a major role in forming and stabilising aggregates (Eq. 2.13).

$$S_t = \frac{SOM}{Clay + Silt\ content} \tag{2.13}$$

Based on analysis of 500 soil samples from semi-arid regions of West Africa, Pieri proposed following three limits of S_t for characterising soil structure:

$S_t \leq 5\%$, loss of soil structure and high susceptibility to erosion,
$S_t = 5\ to\ 7\%$, *unstable* structure and risk of soil degradation,
$S_t \geq 9\%$, *stable* soil structure.

(9) Silica:Sesquioxide Ratio (Tan and Troth 1982).

Nature of clay minerals present in the aggregates has immense effect on development of interparticle bonds, contributing to the stability of soil aggregates, initiating through cations, organic molecules, water and also between clay–clay and clay–coarse grains. Silica:sesquioxide ratio is based on the relative proportion of the cementing agents (sesquioxides or R_2O_3) with the material to be cemented (silica or SiO_2). So, this ratio is obviously a determination of index of erodibility and may range from <1 for nonerodible soils to as high as 9 for erodible soils (Biswas and Nath 1982; Lal and Shukla 2004; Tan and Troth 1982).

Soil colloids with usual size range of 0.2 to 0.005 microns (as high as 2 microns) and humus particles in the millimicrons are the reactive fraction of soil and govern the physical and chemical properties of soil. Types of sesquioxides determine the stability the soil aggregates. Clay particles are better aggregated in the presence of amorphous aluminium hydroxide than that of amorphous iron hydroxide. Water dispersible clay (WDC) can influence water erosion of soil. High clay dispersion ratio (CDR) of soil indicates high erodibility of soil and positively correlated with ($p < 0.51$) dithionite extractable iron (Fed), cation exchange capacity (CEC) and soil organic matter (SOM). Soils with a certain value of percentage of clay dispersion or DR of less than 3 are potentially erodible but such approach cannot distinguish dispersion caused by changes either in electrolytic concentration or other soil properties. In addition, WDC can reduce permeability and crusting of soils. In ferruginous soils of India, water stable aggregates are the result of clay mineral–iron oxide interactions. Water retention in soil matrix against external forces like gravitational forces, centrifugal fields and gradient of external gas pressure is the primary nature of soil in sustaining plant growth in the conditions of water-deficit periods and intermittent supply of water.

The hygroscopic water content can be used as an indicator of clay percentage for reconnaissance soil survey through geophysical signal response. In that context, soil salinity inhibiting plant's ability to take up water from soil is due to osmotic stress, i.e., lowering of the external water potential (Biswas and Mukherjee 1987; Biswas and Nath 1982; dos Santos et al. 2022; Igwe 2005; Pal and Durge 1989, 1993; Pal et al. 2000; Wuddivira et al. 2012; Zhao et al. 2020).

2.6   Soil Loss Causing Vulnerability to Soil Ecosystem—A Major Dialectics of Nature

Soil erosion is the most common problem of land degradation in agricultural, pasture and forest lands. The recent development of the new concept of land degradation neutrality (LDN) requires correct assessment of soil erosion dynamics. The United Nations Convention to Combat Desertification (UNCCD) has defined land degradation neutrality (LDN) as a state of 'amount and quality of land resources, necessary to support ecosystem functions and services and enhance food security, remains stable or increases within specified temporal and spatial scales and ecosystems' (UNCCD 2015). Globally, the state of land degradation neutrality is evaluated through the Sustainable Development Goal (SDG) indicator 15.3 'Proportion of land that is degraded over total land area' and its three subindicators (changes in land cover, land productivity and soil organic carbon) considering the 'one out all out' principle (Akhtar-Schuster et al. 2017; Tsymbarovich et al. 2020; UN 2021).

Water erosion by rainfall and associated fluvial processes and transfer of the mobilised sediment from land surfaces to the oceans through rivers (Fig. 2.6) is an integral part of the natural functioning of the Earth system. This performs as the main driver of formation of current land surface of the earth through the processes of geological cycle of erosion, sedimentation and orogenesis and also the global biogeochemical cycling of carbon, nitrogen and phosphorus as well as numerous other elements. In that way, water erosion plays a key role in developing and maintaining global landscapes and their associated habitats and ecosystems. Fine sediment is frequently referred to as the world's number one pollutant for its role in the transport of many persistent environmental pollutants and, thereby in degrading aquatic habitats through physical and biogeochemical impacts. Very small increases in sediment flux and concentrations can result in lowering of success in hatching and spawning, and thereby reduction in fish populations due to reduction in the availability of dissolved oxygen to the fish eggs during hatching caused by siltation of spawning gravels. On the other hand, reduced sediment loads and river flow may also create problems in ecological balance due to reduction in nutrient supply in some habitats, namely deltas. In that way, erosion and sediment transport have the important role for the sustainable management of the environment and aquatic habitats and ecosystems (Golosov et al. 2019; Kirwan and Megonigal 2013; Lerman and Meybeck 1988; Ludwig et al. 1996; Pinter and Brandon 2005; Tucker and Hancock 2010).

The empirical observation of Howard is supported by the modern scientific observations that the trophic foundation of food webs, formed by soil microbial communities, support terrestrial life through recycling nutrients towards supporting the growth of primary producers and providing the elemental cycling pathways of production and degradation (Howard 1943). Organic biogeochemical by-products of dead and decayed materials from higher trophic levels are changed through soil microbial processes and turned back into the inorganic forms of carbon, nitrogen, phosphorus and other nutrients for growth of plants. In that way, soil productivity works as a proxy for ecosystem health with productive soils supporting the generation of

Fig. 2.6 Sediment transport from land to sea through river: **a** air-view of river system connecting the sea with sediment load; **b** river connecting the sea with sediment load

biomass across trophic levels to meet sustainably the growing demands for agricultural productivity in a healthy ecosystem. In this regard, soil microbial biomass is naturally associated with soil matrix, thereby soil aggregates serve as the functional unit of soil ecosystem (Ciric et al. 2012; Monrozier et al. 1991; Van Gestel et al. 1996; Wilpiszeski et al. 2019).

On the other hand, soil aggregates can perform in stabilising microbial members and enhancing their community interactions. But such soil aggregate and microbial associations can also change community functions and microbial traits through spatial confinement within different-size pores of soil aggregates. Solute diffusion rates are higher in saturated soils, but in drought conditions microorganisms are not able to get dissolved nutrients or signals in pores. As the effective diffusivity is directly correlated with porosity and intrapore geometry, spatial confinement can exert important controls on the rates of nutrient cycling in aggregates, and understanding of metabolic pathways within the broader soil matrix. Soil microbial functional diversity, for example, different aspects of nitrogen cycle, also varies across various sizes of aggregate pore structures. Aggregate structure also controls the hydrological connectivity of soils with profound effect on the microbial community towards isolation of the intra-aggregate communities from one another during dry periods and allowing transport of solutes, metabolites, genetic material and viral particles in wet condition. Therefore, from all those outcomes, it is understandable that each aggregate community may function independently within its local environment for cycling elements and releasing nutrients via metabolic by-products from resident organisms and lysing cells during dry periods. Also, on approaching wet condition soluble carbon is mobilised, thus enabling the flow of metabolites and genetic materials for transfer of new functional capabilities between communities at congenial soil moisture condition. Though soil microbial ecosystems are focused on bulk soils, those are necessarily characterised on microbial communities at the scale of aggregate structures. In that way, soil aggregates represent the relevant length scale of soil pores in shaping the emergent properties of soils and as a whole ecosystem of soil and overall sustainable environmental health (Evans et al. 2016; O'Donnell et al. 2007; Raynaud and Nunan 2014; Schimel 2018; Smith et al. 2017; Vincent et al. 2010; Vontobel et al. 2007; Wilpiszeski et al. 2019).

Soil erosion acts as one of the main threats causing soil degradation across the globe with important impacts on crop yields, soil biota, biogeochemical cycles and ultimately human nutrition. There is consistent decline in soil erosion protection over time across terrestrial biomes, which resulted in systematic increase in global soil erosion rates of 11.7% in between 2006 and 2013 in relation to 2001 as the baseline year. The increase in vegetation cover, central to soil protection, is mostly driven by changes in rainfall erosivity. Apart from that, in the global context, soil erosion has impacted both on the vulnerability of soil conditions and also on soil biodiversity with 6.4% (for soil macrofauna) and 7.6% (for soil fungi) of the vulnerable areas coinciding with regions with high-soil biodiversity. Increase in the amount of vegetation cover leads to a decrease in the risk of water-driven soil erosion under the same environmental and climatic conditions, thus making provision for higher ecosystem service supply. A reasonably accurate estimate and prediction might be helpful in managing soil, but it has limitation for availability of enough mature input data for degradation in the current context of global scale applications of process-based physical models. Such empirical models are applied in dynamically accounting the change in effects of climate and land use land cover due to changes in rainfall erosivity and vegetation cover, respectively (Borrelli et al. 2017; Guerra et al. 2016, 2020; Pimentel et al. 1995).

2.7 Concluding Remarks

Soils in most parts of the world are threatened by erosion. It is a major environmental element and a nonrenewable natural resource. Humans depend on healthy soil to support and regulate the provision of ecosystem services of soil, including provision of food. On the other hand, soil erosion, one of the global environmental problems, can lead to land degradation as well as reducing ecological service levels and on-site impact on biodiversity loss, thereby causing siltation of reservoir/lake and downstream of river. Thus, those processes cause off-site impact on endangering regional sustainable development (Amundson et al. 2015; FAO 2019a; Montanarella 2015; Morgan 2005; Oldeman et al. 1991, 1994; Robinson et al. 2017; Yang et al. 2020).

As one of the global environmental problems, soil erosion possibly accelerates soil carbon loss and increase in atmospheric carbon dioxide level with consequent global impact. Both the actions of the International Symposium on Soil Erosion (GSER19) and the Sixth International Soil Day (Dec 5, 2019) having the common topic of 'stop soil erosion, save our future' aimed at improving soil health by encouraging farmers, social groups and governments to actively participate in soil erosion control (Crosson et al. 1995; FAO 2019b; Lal 2019; Montanarella 2015; Yang et al. 2020).

Based on a wide range of statistical, remote sensing, mathematical modelling data, obtained on the basis of scientific and field studies performed at different levels, it is asserted that in Russia the total area of eroded lands and those under erosion risk occupy more than half of all agricultural lands, whereas during 1950s to 1980s soil fertility of croplands decreased by 30–60% only due to water erosion. On the other hand, abandonment of arable land and subsequent overgrown with natural vegetation during the last 30–40 years indicate a decrease in erosion rate even in the area of eroded land. There in Russia the decrease in depth of soil freezing and flow of spring runoff due to climate change have resulted in decrease of soil erosion. According to the 'one out all out principle', three global land degradation neutrality (LDN) indicators are soil organic carbon (SOC), net primary productivity (NPP) and land cover change. Soil erosion (SE) indicator can be used as complement to those three global LDN indicators. SE as 'Rate of soil loss' (ton ha^{-1} yr^{-1}) and 'Total soil loss' (1000 tons, in certain area during selected time period can be interpreted at national and subnational levels, respectively, with wider and site-specific data, including those obtained through remote sensing data based on the classifier of thematic applications of remote sensing technologies on the basis of local site observations (Tsymbarovich et al. 2020; UNCCD 2016).

Since the 1990s, a series of studies on assessing and mapping soil erosion have been carried out giving rise to valuable data sets on pattern of soil erosion and hotspots of the main drivers of soil erosion in some countries at regional scales with great significance in supporting soil conservation policies. Those works are mostly outcomes of low- and medium-resolution remote sensing imagery and small carto-graphic scale in geographic information system (GIS) investigations, and thus, those data cannot fully meet the requirements of a comprehensive understanding of soil erosion and related decision-making. Therefore, soil erosion assessment should be based on field investigation, high-resolution remote sensing and/or detailed carto-graphic data, and the soil erosion information finally obtained at the individual field scale. That information will be helpful in planning for suitable soil and water conser-vation measures at farmers' field level (Borrelli et al. 2017, 2021; Crosson et al. 1995; FAO 2019b; Oldeman et al. 1991, 1994; Teng et al. 2018; Yang et al. 2020).

References

Akhtar-Schuster M, Stringer LC, Erlewein A, Metternicht G, Minelli S, Safriel U, Sommer S (2017) Unpacking the concept of land degradation neutrality and addressing its operation through the Rio conventions. J Environ Manag 195(1):4–15. https://doi.org/10.1016/j.jenvman.2016.09.044

Amundson R, Berhe AA, Hopmans J W, Olson C, Sztein AE, Sparks DL (2015) Soil and human security in the 21st century. Sci 348(6235):1261071–1261071e1261071–1261076

Anderson HW (1954) Suspended sediment discharge as related to stream flow topography, soil and land use. Trans Am Geophys Union 35:268–281

Baver LD (1935) Factors contributing to the genesis of soil micro-structure. Amer Soil Surv Assoc Bul 18:55–56

Bennett HH (1939) Soil conservation. McGraw-Hill, New York and London

Biswas TD, Mukherjee S (1987) Textbook of soil science. Tata McGraw-Hill Publishing Co. Ltd., New Delhi, p 84

Biswas TD, Nath J (1982) Clay mineralogy and soil physical behaviour. In: Review of soil research in India, Part II, transactions XII international congress soil science, pp 740–745

Blanco-Canqui H, Lal R (2008) Principles of soil conservation and management. Springer, New York, pp 42–45. https://doi.org/10.1007/978-1-4020-8709-7

Borrelli P, Robinson DA, Fleischer LR, Lugato E, Ballabio C, Alewell C, Meusburger K, Modugno S, Schütt B, Ferro V, Bagarello V, Van Oost K, Montanarella L, Panagos P (2017) An assessment of the global impact of 21st century land use change on soil erosion. Nat Comm 8(1):1–13

Borrelli P, Alewell C, Alvarez P, Anache JAA, Baartman J, Ballabio C, Bezak N, Biddoccu M, Cerdà A, Chalise D, Chen S, Chen W, De Girolamo AM, Gessesse GD, Deumlich D, Diodato N, Efthimiou N, Erpul G, Fiener P, Freppaz M, Gentile F, Gericke A, Haregeweyn N, Hu B, Jeanneau A, Kaffas K, Kiani-Harchegani M, Lizaga Villuendas I, Li C, Lombardo L, López-Vicente M, Lucas-Borja ME, Märker M, Matthews F, Miao C, Mikoš M, Modugno S, Möller M, Naipal V, Nearing M, Owusu S, Panday S, Patault E, Patriche CV, Poggio L, Portes R, Quijano L, Reza Rahdari M, Renima M, Ricci GF, Rodrigo-Comino J, Saia S, Nazari Samani A, Schillaci C, Syrris V, Kim HS, Noses Spinola D, Tarso Oliveira P, Teng H, Thapa R, Vantas K, Vieira D, Yang JE, Yin S, Zema DA, Zhao G, Panagos P (2021) Soil erosion modelling: a global review and statistical analysis. Sci Total Environ 780:146494

Bouyoucos GJ (1935) The clay ratio as a criterion of susceptibility of soils to erosion. J Am Soc Agron 27(9):738–741

Briggs LJ, McLane JW (1907) The moisture equivalents of soils. USDA Bureau Soils Bull 45:1907

Briggs LJ, Shantz HL (1912) The coefficient for different plants and its indirect determination. USDA Bur Plant Indus 230:1–82

Chepil WS, Woodruff NP (1963) The physics of wind erosion. Adv Agron 15:211–302

Chorley RJ (1959) The geomorphic significance of some Oxford soils. Am J Sci 257:503–515

Ciric V, Manojlovic M, Nesic L, Belic M (2012) Soil dry aggregate size distribution: effects of soil type and land use. J Soil Sci Plant Nutr 12:689–703. https://doi.org/10.4067/S0718-951620120 05000025

Crosson P, Pimentel D, Harvey C et al (1995) Soil erosion estimates and costs. Science 269(5223):461–465

Cruse RM, Hajek BF, Karlen DL, Lowery B, Power JF, Schumacher TE, Skidmore EL, Sojka RE, Arnold RW, Bauer A, Brownfield SH, Bruce RR, Culley JLB, Dowdy RH, Dunnigan LP, Ewing RP, Fox HD, Gupta SC, Mielke LN, Mulla DJ, Rads WJ, Romkens MJM, Schmidt BL, Singer MJ, Smith SJ, Soileau JM, Van Doren CA, Zobeck TM (1990) Erosion and soil properties. In: Soil erosion and soil productivity, pp 23–39. Available https://infosys.ars.usda.gov/WindErosion/publications/Andrew_pdf/89-421-A.pdf. Accessed 18 June 2022

Dakshinamurti C, Gupta RP (1968) Practicals in soil physics. Indian Agric Res Inst 29:60

dos Santos TB, Ribas AF, de Souza SGH, Budzinski IGF, Domingues DS (2022) Physiological responses to drought, salinity, and heat stress in plants: a review. Stresses 2:113–135. https://doi.org/10.3390/stresses2010009

Einstein HA, El-Samni EA (1949) Hydrodynamic forces on rough wall. Rev Mod Phys 21(3):520–524

European Union (2014) Study on soil and water in a changing environment. Final Report, Annex 3, Table 5, p 62. https://doi.org/10.2779/20608

Evans S, Dieckmann U, Franklin O, Kaiser C (2016) Synergistic effects of diffusion and microbial physiology reproduce the birch effect in a micro-scale model. Soil Biol Biochem 93:28–37. https://doi.org/10.1016/j.soilbio.2015.10.020

FAO (2000) Land resource potential and constraints at regional and country levels. World soil resources report 90. Land and water development division, food and agriculture organization of the United Nations (FAO), Rome

FAO (2019a) Soil erosion: the greatest challenge to sustainable soil management. Food and agriculture organization of the United Nations (FAO), Rome

FAO (2019b) Outcome document of the global symposium on soil erosion. Food and agriculture organization of the United Nations (FAO), Rome

Flanagan DC, Ascough JC II, Nieber JL, Misra D, Douglas-Mankin KR (2013) Advances in soil erosion research: processes, measurement and modelling. Trans Am Soc Agril Biol Eng 56(2):455–463

Gao B, Walter MT, Steenhuis TS, Parlange JY, Richards BK, Hogarth WL, Rose CW (2005) Investigating raindrop effects on transport of sediment and non-sorbed chemicals from soil to surface runoff. J Hydrol 308:313–320

Glinski J, Horabik J, Lipiec J (2011) Encyclopedia of agrophysics. Springer, Dordrecht

Golosov V, Walling DE, Mishra A (2019) Erosion and sediment problems: global hotspots. United Nations educational, scientific and cultural organization (UNESCO), Paris, pp 8–14

Guerra C, Maes J, Geijzendorffer I, Metzger MJ (2016) An assessment of soil erosion prevention by vegetation in Mediterranean Europe: current trends of ecosystem service provision. Ecol Ind 60:213–222

Guerra CA, Rosa IMD, Valentini E, Wolf F, Filipponi F, Karger DN, Xuan AN, Mathieu J, Lavelle P, Eisenhauer N (2020) Global vulnerability of soil ecosystems to erosion. Landsc Ecol 35:823–842. https://doi.org/10.1007/s10980-020-00984-z

Hénin S, Monnier G, Combeau A (1958) Methode pour l'etude de la stabilite structural des sols. Ann Agron 9:73–92

Howard A (1943) An agricultural testament. Oxford University Press, New York

Hudson N (1981) Soil conservation, 2nd edn. Cornell State University, New York

Igwe CA (2005) Erodibility in relation to water-dispersible clay for some soils of eastern Nigeria. Land Degrad Develop 16:87–96

Ippen AT, Verma RP (1953) Reprinted from processing Minnesota intern hydraulic conversion, Hydrodynamics Lab, Massachusetts institute technical, Cambridge

Kay BD, Angers DA, Groenovelt PH, Baldock JA (1988) Quantifying the influence of cropping history on soil structure. Can J Soil Sci 68:359–368

Kirwan ML, Megonigal JP (2013) Tidal wetland stability in the face of human impacts and sea level rise. Nature 504:53–60

Lal R (2019) Accelerated soil erosion as a source of atmospheric CO_2. Soil Tillage Res 188:35–40. https://doi.org/10.1016/j.still.2018.02.001

Lal B, Biswas H, Raj S (2006) Soil erodibility characteristics under different land uses in Bundelkhand. Indian J Agrofor 8(1):69–73

Lal R, Shukla MK (2004) Principles of soil physics. Marcel Dekker, Inc., New York, pp 121–125

Lerman A, Meybeck M (eds) (1988) Physical and chemical weathering in geochemical cycles. Springer, Netherlands

Lu J, Zheng F, Li G, Bian F, An J (2016) The effects of raindrop impact and runoff detachment on hillslope soil erosion and soil aggregate loss in the Mollisol region of Northeast China. Soil Tillage Res 161:79–85. https://doi.org/10.1016/j.still.2016.04.002

Ludwig W, Probst J-L, Kempe S (1996) Predicting the oceanic input of organic carbon by continental erosion. Glob Biogeochem Cycles 10:23–41

Ma RM, Li ZX, Cai CF, Wang J-G (2014) The dynamic response of splash erosion to aggregate mechanical breakdown through rainfall simulation events in Ultisols (Subtropical China). CATENA 121:279–287. https://doi.org/10.1016/j.catena.2014.05.028

Middleton HE, Slater CS, Byres HG (1932) Physical and chemical characteristics of the soils from the erosion experiment stations. USDA Tech Bull 316

Middleton HE, Slater CS, Byres HG (1934) The physical and chemical characteristics of the soils from the erosion experiment stations—second report. USDA Tech Bull 430

Middleton HE (1930) Properties of soil which influence soil erosion. United States department of agriculture (USDA) Tech Bul 178

Monrozier LJ, Ladd JN, Fitzpatrick RW, Foster RC, Raupach M (1991) Components and microbial biomass content of size fractions in soils of contrasting aggregation. Geoderma 50:37–62. https://doi.org/10.1016/0016-7061(91)90025-O

Montanarella L (2015) Govern our soils. Nature 287(7580):32–33

Morgan RPC (2005) Soil erosion and conservation, 3rd edn. Blackwell Science Ltd, Malden

Myers N (1993) Gaia: an atlas of planet management. Anchor/Doubleday. Garden City, New York

O'Donnell AG, Young IM, Rushton SP, Shirley MD, Crawford JW (2007) Visualization, modelling and prediction in soil microbiology. Nat Rev Microbiol 5:689–699. https://doi.org/10.1038/nrmicro1714

Oldeman LR, Hakkeling TTA, Sombroek WG (1994) The global extent of soil degradation. In: Green D, Szabolcs J (eds) Soil resilience and sustainable land use. CAB International, Wallingford, pp 99–118

Oldeman LR, Hakkeling RTA, Sombroek WG (1991) World map of the status of human-induced soil degradation: An explanatory note, 2nd revised edition. Global assessment of soils degradation, The map sheets. World soil information (ISRIC), Wageninen

Pal DK, Durge SL (1989) Release and adsorption of potassium in some benchmark alluvial soils of India in relation to their mineralogy. Pedol 39:235–248

Pal DK, Durge SL (1993) Potassium release from clay micas. J Indian Soc Soil Sci 41:67–69

Pal DK, Bhattacharyya T, Deshpande SB, Sarma VAK, Velayutham M (2000) Significance of minerals in soil environment of India. NBSS review series 1, NBSS & LUP, Nagpur, India, pp 19–48

Pieri C (1991) Fertility of soils: a future for farming in the West African savannah. Springer, Berlin

Pimental D, Terhune EC, Dyson-Hudson R, Rocherau S (1976) Land degradation: effects on food and energy resources. Sc 94:149–155

Pimentel D, Harvey C, Resosudarmo P, Sinclair K, Kurz D, McNair M, Crist S, Shpritz L, Fitton L, Saffouri R, Blair R (1995) Soil erosion estimates and costs. Sci 267(5201):1117–1123. https://doi.org/10.1126/science.267.5201.1117

Pinter N, Brandon MJ (2005) Erosion builds mountains. Sci Am 15:74–81

Raynaud X, Nunan N (2014) Spatial ecology of bacteria at the microscale in soil. PLoS ONE 9:e87217. https://doi.org/10.1371/journal.pone.0087217

Robinson DA, Panagos P, Borrelli P (2017) Soil natural capital in Europe; a framework for state and change assessment. Sci Rep 7(1):6706

Schimel JP (2018) Life in dry soils: effects of drought on soil microbial communities and processes. Annu Rev Ecol Evol Syst 49:409–432. https://doi.org/10.1146/annurev-ecolsys-110617-062614

Singh RA (1980) Soil physical analysis, 2nd edn. Kalyani Publishers, New Delhi, pp 39–40

Smith AP, Bond-Lamberty B, Benscoter BW, Tfaily MM, Hinkle CR, Liu C, Bailey VL (2017) Shifts in pore connectivity from precipitation versus groundwater rewetting increases soil carbon loss after drought. Nat Commun 8:1335. https://doi.org/10.1038/s41467-017-01320-x

Tan KH, Troth PS (1982) Silica-sesquioxide ratios as aids in characterization of some temperate region and tropical soil clays. Soil Sci Soc Am J 46(5):1109–1114. https://doi.org/10.2136/sssaj1982.0361599500460005004

Teng H, Liang Z, Chen S, Liu Y, Raphael A, Rossel V, Chappell A, Wu Y, Shi Z (2018) Current and future assessments of soil erosion by water on the Tibetan plateau based on RUSLE and CMIP5 climate models. Sci Total Environ 635:673–686. https://doi.org/10.1016/j.scitotenv.2018.04.146

Tsymbarovich P, Kust G, Kumani M, Golosov V, Andreeva O (2020) Soil erosion: an important indicator for the assessment of land degradation neutrality in Russia. Int Soil Water Conserv Res 8(4):418–429. https://doi.org/10.1016/j.iswcr.2020.06.002

Tucker GE, Hancock GR (2010) Modelling landscape evolution. Earth Surf Proc Landf 35:28–50. https://doi.org/10.1002/esp.1952

UN (2021) The sustainable development goals report. The sustainable development goals website of the United Nations, department of economic and social affairs statistics division, United Nations (UN), New York. Available https://unstats.un.org/sdgs/report/2021/. Accessed 18 June 2022

UNCCD (2015) ICCD/COP (12)/4. Report of the IWG on LDN. Integration of the sustainable development goals and targets into the implementation of the UNCCD. In: Conference of the parties, Twelfth session, Ankara, Turkey. United Nations convention to Combat Desertification (UNCCD), Bonn

UNCCD (2016) Monitoring land degradation neutrality status. United Nations convention to Combat Desertification (UNCCD) Knowledge Hub. UNCCD Secretariat, Bonn Available https://knowledge.unccd.int/knowledge-products-and-pillars/guide-scientific-conceptual-framework-ldn/key-elements-scientific-5. Accessed 18 June 2022

Van Gestel M, Merckx R, Vlassak K (1996) Spatial distribution of microbial biomass in microaggregates of a silty-loam soil and the relation with the resistance of microorganisms to soil drying. Soil Biol Biochem 28:503–510. https://doi.org/10.1016/0038-0717(95)00192-1

Veihmeyer FJ, Hendrickson AH (1931) The moisture equivalent as a measure of field capacity of soils. Soil Sci XXXIII 3:181–194. https://doi.org/10.1097/00010694-193109000-00003

Vincent ME, Liu W, Haney EB, Ismagilov RF (2010) Microfluidic stochastic confinement enhances analysis of rare cells by isolating cells and creating high density environments for control of diffusible signals. Chem Soc Rev 39:974–984. https://doi.org/10.1039/b917851a

Vontobel P, Lehmann E, Laloui L, Vulliet L, Flühler H (2007) Water flow between soil aggregates. Transp Porous Media 68:219–236. https://doi.org/10.1007/s11242-006-9041-z

Wilpiszeski RL, Aufrecht JA, Retterer ST, Sullivan MB, Graham DE, Pierce EM, Zablocki OD, Palumbo AV, Elias DA (2019) Soil aggregate microbial communities: towards understanding microbiome interactions at biologically relevant scales. Appl Environ Microbiol 85(14):e00324-e419. https://doi.org/10.1128/AEM.00324-19

Wischmeier WH, Mannering JV (1969) Soil and water management and conservation. Relation of
  soil properties to its erodibility. Proc Soil Sci Soc Am 33:131–137. https://doi.org/10.2136/sss
  aj1969.03615995003300010035x
Wuddivira MN, Robinson DA, Lebron I, Brèchet L, Atwell M, De Caires S, Oatham M, Jones SB,
  Abdu H, Verma AK, Tuller M (2012) Estimation of soil clay content from hygroscopic water
  content measurements. Soil Sci Soc Am J 76(5):1529–1535. https://doi.org/10.2136/sssaj2012.
  0034
Wuest SB, Williams JD, Gollany HT (2006) Tillage and perennial grass effects onponded infiltration
  for seven semi-arid loess soils. J Soil Water Conserv 61(4):218–222
Yang Q, Zhu M, Wang C, Zhang X, Liu B, Wei X, Pang G, Du C, Yang L (2020) Study on a soil
  erosion sampling survey in the pan-third pole region based on higher-resolution images. Int Soil
  Water Conserv Res 8:440e451
Young IM, Crawford JW, Rappoldt C (2001) New methods and models forcharacterising structural
  heterogeneity of soil. Soil Tillage Res 61:33–45
Zhao C, Zhang H, Song C, Zhu J-K, Shabala S (2020) Mechanisms of plant responses and adaptation
  to soil salinity. Innov 1(1):100017. https://doi.org/10.1016/j.xinn.2020.100017
Zobeck TM, Baddock M, Pelt RSV, Tatarko J, Acosta-Martinez V (2013) Soil property effects on
  wind erosion of organic soils. Aeolian Res 10:43–51

Chapter 3
Impact of Climate, Water and Biological Factors on Soil Health

Abstract Impact of climate change and land use management on soil health can be assessed by some indicators like aggregate stability, soil organic matter, carbon and nitrogen cycling, microbial biomass and activity and microbial fauna and flora diversity. For that purpose, a minimum data set is important for useful quantitative application of soil health concept and starting suitable soil and water conservation measures on farm land, especially on individual plot for assuring sustainable crop cultivation and food security.

Keywords Carbon · Climate change · Groundwater · Land · Management · Soil health · Soil organic matter

Abbreviations

AWC	Available water-holding capacity
BD	Bulk density
C	Carbon
CO_2	Carbon dioxide
CSA	Climate smart agriculture
DEFRA	Department for Environment, Food & Rural Affairs
EC	Electrical conductivity
FAO	Food and Agriculture Organization of the United Nations
GISTEMP	GISS Surface Temperature Analysis
H_2CO_3	Carbonic acid
H^+	Proton
IPCC	Intergovernmental Panel on Climate Change
ITPS	Intergovernmental Technical Panel on Soils
K	Potassium
MBC	Microbial biomass carbon
Mg	Magnesium
N	Nitrogen
P	Phosphorus

© The Author(s), under exclusive license to Springer Nature Switzerland AG 2022
S. Panda, *Soil and Water Conservation for Sustainable Food Production*,
Chemistry of Foods, https://doi.org/10.1007/978-3-031-15405-8_3

PCA   Principal component analysis
PPP   Plant primary productivity
PMN   Potentially mineralisable nitrogen
POM   Particulate organic matter
SAR   Sodium adsorption ratio
SOM   Soil organic matter
SQE   Soil quality element
SQE1   Food and fibre production
SQE2   Erosivity
SQE3   Groundwater quality
SQE4   Surface water quality
SQE5   Air quality
SQE6   Food quality
SQI   Soil quality index

3.1 Introduction

Soil health is defined as 'the capacity of soil to function as a vital living system, within ecosystem and land-use boundaries, to sustain plant and animal productivity, maintain or enhance water and air quality, and promote plant and animal health' (Doran and Zeiss 2000).

Soil is the part and parcel of hydrological cycle, responsible for global water cycle influencing climate regulation and nutrient cycles for existence of all life forms. Soil health indicators can be regarded as a composite set of physical, chemical and biological attributes as related to soil functions congenial for healthy soils which, in turn, are affected by management and climate change drivers. Soil health is the expression of impacts of global change drivers like rising level of atmospheric carbon dioxide (CO_2), aberrant rainfall (or precipitation) pattern and deposition of atmospheric nitrogen on three attributes of soil functions in relation to climate change. Mankind depends on agriculture for food and other important livelihood requirements. Agriculture covers about 38% of land surface with soils affording about 95% of global food production. Finding out changes in soil health (or soil quality) properties in relation to predicted climate changes is necessarily required to understand and to act accordingly for effective management of the changes in soil properties for farming without worries for sustainable food production in the present context of climate change (Borrelli et al. 2020; French et al. 2009; Montanari et al. 2013). Principles of soil health include: (1) soil surface protection with growing plants or mulch, (2) soil profile disturbance as little as possible, (3) increase in diversity of plant species through crop rotations, cover crops and polycultures, (4) soil biota enrichment, (5) soil organic matter (SOM) built-up and (6) possible integration of livestock in farming systems. Application of those soil health principles advances key functions of soils in agroecosystems, viz. water availability and storage, nutrient availability and cycling, soil structural stability, root growth, carbon (C) storage, biodiversity conservation and filtering/ buffering of potential water pollutants (Al-Kaisi and Lal 2017; Brussaard et al. 2007; Keesstra et al. 2012; Kibblewhite et al. 2008; Mehra et al. 2018; Nannipieri et al. 2003). Soil and water resources have now become the most endangered natural resources in the present alarming conditions of land degradation and climate change. According to FAO and ITPS, globally an estimated 20–40 billion tons of fertile soils are lost every year. So, it requires action for setting up sustainable land management as a common goal for farming and land management for improvement of soil health to conserve soil and water. Soil is the major reserve of C in terrestrial ecosystems, and soil C plays the role of C-sequestration and thereby plays the important role in global soil C cycle and climate change and simultaneously manoeuvring sustainability in soil health and productivity. Globally, climate change and climate variability are two risks for farming. According to the Intergovernmental Panel on Climate Change (IPCC), climate change is the long-term (decadal or normal) change in the mean and/or variability of climate attributes, and in contrast, climate variability is the short-term (a day, month, season or years) fluctuations in temperature, rainfall distribution and climate variables. Scientists are now certain that environmental changes (i.e., climate change, land degradation, biodiversity loss, etc.) are seriously affecting soil health and agricultural productivity. Rising population trend (~9.8 billion by 2050) in conjunction with 125% increase in global temperature anomaly (0.44 °C in 2000 to 0.99 °C in 2016) due to global warming will impede food self-sufficiency and security of many nations, having source of food production from a minute fraction (around 10.9%, i.e., 1.4 billion ha out of ~13.4 billion ha of total world area) of land used for food production (FAO 2016; FAO and ITPS 2015; GISTEMP Team 2018; IPCC 2012; Mehra et al. 2018; Potter et al. 2017; Singh et al. 2011, 2018a; Thornton et al. 2014; Titeux et al. 2016).

Soil water retention provides multiple ecosystem services of mainly three major functions like (1) provisioning, (2) regulating and (3) supporting services for sustaining human needs involving environment, health and socio-economic aspects. (1) Provisioning services makes the ecosystem able to produce food, feed and fibre including provision for drinking water through surface storage and groundwater recharge, prevention of groundwater contamination from surface pollution through filtering action of soil and its profile; (2) regulating services modulates climate through the processes of evapotranspiration and flood and drought control; and (3) supporting services through regulation of soil water conditions like erosion control becomes necessary for safeguarding soil functions active in the long run. In that way, supporting services are necessary for ensuring long-term benefits necessary for the production of all other ecosystem services, e.g., supporting the presence of water within soil for regulating nutrient cycling; supporting biodiversity build-up for biomass production and water filtration (Brauman et al. 2007; Dominati et al. 2010; European Union 2014; MEA 2005).

3.2   Formulation of Soil Quality Indexing Including Water Quality in the Context of Climate Change

Soil health characterises scores of soil functions and ecosystem services of net primary production (NPP), filtering and denaturing of contaminants for water purification, improving air quality by removing pollutants, improving environment and modifying local, regional and global climates. Similar key soil function parameters (or indicators) are used for assessing both soil quality and soil health. The results are characterised quantitatively for soil quality and qualitatively for soil health, while the latter term gives more emphasis on ecological functions and soil biodiversity which substantiate natural perspectives of soil functioning as dynamic living resource and with capacity for self-organisation including resistance to soil degradation and ability to recover against a perturbation due to inherent resilience in soil system (Allen et al. 2011; Herrick et al. 1999; Karlen et al. 2003; Lal 2011; Magdoff 2001).

For monitoring climate change-related modifications in soil health functions, evaluation of the soil health indicators is becoming difficult due to paucity in existing data as obtained through the rise in integrative process in soil health tests. For that purpose, threshold level of soil function indicators is generally considered. This is helpful for indirect estimation of soil function, for tracing out the present trend in soil health for assessing environmental impacts and soil health as well with time for monitoring changes in soil properties through linking all possible interrelationships among soil health attributes towards sustainable land management. Evaluation of soil health process includes a series of actions, viz. (i) selection of soil health indicators, (2) confirmation of minimum data set, (3) generation of process of interpretation of indices and (4) validation through on-farm assessment. Sustainable agricultural productivity and ecosystem functionality, in general, are affected by elevated atmospheric CO_2 concentration, increasing atmospheric temperature, atmospheric N deposition, fluctuation in total and seasonal rainfall pattern and occurrences of extreme climatic events like droughts, floods, etc. On the other hand, those factors affect soil biological processes and C and N cycling with the consequences on soil structure, erosion, plant nutrient availability and plant diseases which have direct effect on sustainability in food production (Allen et al. 2011; Dalal et al. 2003a, b; Doran 2002; Doran and Zeiss 2000; Idowu et al. 2009; Kinyangi 2007; Riley 2001).

Doran and Parkin put forward a strategy to assess a soil quality (SQ) as shown in Eq. (3.1) (Doran and Parkin 1994):

$$SQ = f(SQE1, SQE2, SQE3, SQE4, SQE5, SQE6) \qquad (3.1)$$

where the specific soil quality elements (SQEi) are defined as follows:

(a) Food and fibre production = SQE1
(b) Erosivity = SQE2
(c) Groundwater quality = SQE3
(d) Surface water quality = SQE4
(e) Air quality = SQE5

(f) Food quality = SQE6.

For the purpose of soil quality monitoring, instead of interpreting momentary data, soil quality index (SQI) approach is useful over time, though it cannot be directly related to any specific function or indicator for categorising high- or low-quality index, because individualities of all functions and indicators are compounded through the process of indexing. Though soil quality approach is not useful in all parts of the land where erosion is the main menace, it requires immediate action for minimising soil erosion. Still, adverse effects of soil compaction and nutrient leaching have attracted all farming communities, in general, for maintaining sustainability in food production. There are various efforts to alleviate complexities in representation of functional structure approach for easier interpretations through use of multivariate tool like principal component analysis (PCA) and factor analysis. Such attempts are recognised as variable reduction techniques, and, thus, those multivariate tools are useful to find out high correspondence between a certain indicator and a component for obtaining certain absolute loading. Still, such approach is unable to resolve contradiction between two different soil functions, e.g., high productivity requiring high rate of N mineralisation, which in turn may be risky for environment from nutrient contamination by leaching, whereas hard clay pan may prevent such nutrient leaching, but at the same time that may depress root growth and thereby food production. But microorganisms, with dynamic contact with soil matrix, are regarded as potential indicators of soil quality and as regulator of soil quality changes. Soil microbial biomass is part of the active ingredient in soil for nutrient cycling, degradation of organic pollutants and slowing down rate of dissolution of pollutants in water through building of organo-pollutant complex (Doran et al. 1996; Gonzales-Prieto et al. 1992; Nolin et al. 1989; Stenberg 1999; Stenberg et al. 1998).

On reviewing a range of frameworks of identifying soil health indicators (Dalal and Moloney 2000; Dalal et al. 1999; Doran 2002; Kibblewhite et al. 2008; Lal 1999; Nuttall 2007; Schjønning et al. 2004; Stenberg 1999) to measure change and implementing strategies to cope up with the climate change, Allen and coworkers have classified indicators of soil health functions into three major attributes of soil physical, chemical and biological properties, as shown in Fig. 3.1 (Allen et al. 2011).

Indexing soil health indicators require evaluation of a set of soil health quality indicators for calculation of an index useful for identification of problem soils, establishment of land values, monitoring influence of agricultural management, etc., and which would be applicable for securing sustainable food production.

3.3 Effect of Water on Soil Health in the Context of Climate Change

Water availability and storage are key soil functions in agroecosystems. Those ensure surface water storage, groundwater recharge and root growth and thereby enhance

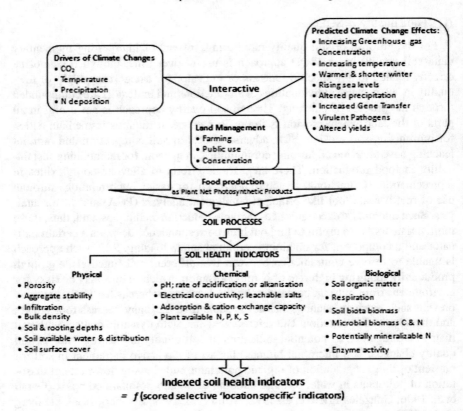

Fig. 3.1 Estimation of soil health index involving drivers of climate change, predicted climate change effects and land management, food production, soil processes and soil health indicators with location specific index of soil health (following Allen et al. 2011)

soil biodiversity helpful for provisioning sustainable food production. Water availability and storage also help for ensuring decontamination of polluted water for supply of fresh water, resistance against landscape degradation, etc., towards better environment. Soil physical health is dependent on management and ambient climate factors like management of SOM, soil structure, bulk density, air and water movement through soil profile. Movement and management of water, oxygen and nutrients influence one of the most important attributes of a healthy soil, e.g., soil structural stability. Imbalance in two dynamic soil properties of soil structure and SOM can significantly impact water retention and susceptibility of soil to erosion. Movement of gas and water through the soil profile also affects processes like evaporation, heat conduction and convective transfer, and, thus, ultimately those soil functions control soil-thermal regime. In that way, rise in air and soil temperature causes imbalance in soil regime and enhances key soil processes of rapid decomposition of SOM and volatilisation loss of nutrient (Basher and Ross 2002; Christensen et al. 1996; Datiri and Lowery 1991; DEFRA 2005; Głab and Kulig 2008; Keesstra et al. 2012; Kibble-white et al. 2008; Kladivko 2001; Lal 2003; Luo et al. 2017; Mehra et al. 2018; Nannipieri et al. 2003; Qian et al. 2011; Rogers and McCarty 2000; Salem et al. 2015; Sojka et al. 2007; Ussiri et al. 2009).

In the wake of 'Dust Bowl era' of 1930s, numerous studies have reported that no-till improves water conservation and is favourable for soil-thermal regime in the topsoil surface (Blevins et al. 1983; Ehlers and Claupein 1994; Merrill et al. 1996). Recent advances in tillage systems such as precision tillage and strategic tillage are able to optimise edaphic conditions, e.g., soil–water and soil–temperature regimes, porosity, soil aeration, seed-soil contact and nutrient availability. Climate change has also impact on soil chemical health; e.g., changes in precipitation may influence the extent of soil acidification (Charman and Murphy 2007; Wild 1993). Rainwater, which is usually slightly acidic (pH: 5–5.6), may release protons (H^+) in soil from the dissolution of CO_2 and dissociation of carbonic acid (H_2CO_3), though this process is unlikely to affect soil pH on its own (because H_2CO_3 is a weak acid). Acid soils are common in the wet tropics, because enhanced leaching of basic cations to subsoil horizons facilitates the process of soil acidification (Rengel 2011). Though soil acidification being a natural process, nitrate leaching becomes the main cause of soil acidification during high rainfall conditions (Bolan et al. 1991). In addition to rainfall, soil moisture content can be affected by the high specific heat and low diffusivity of soil water and also other constituents of soil such as clay and SOM. Nutrient uptake by plant roots from soil takes place through soil water, which is decreased in dry soil at short distances (e.g., diffusion of phosphate and zinc), as well as the mass flow of water-soluble nutrients like sulphates, nitrate, calcium, magnesium and silicon at larger distances (Barber 1996; Mackay and Barber 1985; Monteith and Unsworth 2013). Lal reported that crop yields decrease exponentially with increase in aridity as the shortage of available soil moisture can also influence the availability and transportation of soil nutrients to the plant (Lal 2000). Under optimum moisture conditions, the increase of root zone temperature from 15–18 °C to 25–29 °C enhanced the nutrient (e.g., K and P) uptake up to 100–300%, by increasing the root surface area, as well as increase in rate of nutrient diffusion and water influx (Brouder and Volenec 2008; Ching and Barber 1979; Mackay and Barber 1984). It is also to be noted that soil fauna mix with soil particles thereby makes pores, tunnels and other biological compartments, air and water flow through soil, thus further the soil microbial activity (Lavelle et al. 2006; Mehra et al. 2018).

Inclusion of diverse crop rotations and cover crop mixes in farming systems can improve soil cover, biodiversity and ecosystem functions, such as plant available nutrients, water storage, nutrient use efficiency and crop yield (Lorenz and Lal 2014; Poeplau and Don 2015; Sarker et al. 2018a, b). While deforestation provides wood products to meet the needs of a growing population, there are negative effects of this land use change on ecosystem services such as increase in soil erosion (wind and water), decline of water regulation against floods and landslides, loss of plant biodiversity and a decline in aesthetic and recreational values. Due to unsustainable land use and management, ~24% of global land had already suffered from declines in soil health and plant productivity over the past quarter of a century (Rhodes 2014). It is reported that conservation tillage practices, especially when combined with stubble retention, may increase SOC (Mangalassery et al. 2014; Petersen et al. 2008), water retention (via increasing infiltration and minimising runoff), nutrient use efficiency (Triplett and Dick 2008) and herbicide efficacy (Bajwa et al. 2017). Mehra and colleagues have noted that in addition to many other tangible soil health benefits, no-till contributes to mitigation of climate change through reducing the oxidation rate of SOM and per hectare fuel consumption (Mehra et al. 2018), thus attracting farmers' interest towards no-till farming which is, thus, turned to be a mainstream approach in agricultural systems globally (Lal 2004; Singh et al. 2018).

The evidence from empirical studies suggests that there are four main areas that require support to facilitate adoption of CSA (climate-smart agriculture) which are as follows: (1) dissemination/extension of information and particularly to adapt practices to local conditions; (2) coordination of efforts for objectives best managed at landscape scale, such as reducing the risks of flood, managing pest outbreaks or preserving biodiversity; (3) comprehensive risk assessment strategies to improve climate change resilience; and (4) reliable and timely access to specific inputs in farming systems to support resource-use efficiency. Recognising the cross-cutting nature of these requirements, it is, therefore, needed to focus on the nexus between soil health and climate change to address challenges of meeting global needs for food, fresh water and biomass-based products. The conceptual framework of this revolution should focus on improving soil health and mitigating climate impacts via the adoption of CSA practices, which is one of the central pillars to address the climate–food–water nexus. New developments in plant breeding have enhanced the resilience of crop varieties to drought and other climate extremes such as high temperature and waterlogging. The use of mixed cropping (polycultures) with several crops growing at one time can increase resilience during periods of water or heat stress. In this regard, promising systems include integrated natural resource management, i.e., integrated management of land, water and biological resources (Kole et al. 2015; Mehra et al. 2018; Misra 2014).

'The capacity of the soil to function' can be represented by important six soil functions such as (i) water flow and retention; (ii) solute transport and retention; (iii) physical stability and support; (iv) retention and cycling of nutrients; (v) buffering and filtering of potentially toxic materials; and (vi) maintenance of biodiversity and habitat (Daily et al. 1997; Karlen et al. 1997). Based on broader ecological-based approach, Doran and coworkers have defined soil health as 'the continued capacity of soil to function as a vital living system, within ecosystem and land-use boundaries, to sustain biological productivity, maintain the quality of air and water environments, and promote plant, animal, and human health' (Doran et al. 1996). Frequently recommended soil quality indicators include (a) available water-holding capacity (AWC) and (b) infiltration rate (Doran and Parkin 1996; Karlen et al. 1997; Wienhold et al. 2009) along with other twelve soil quality indicators of SOM, particulate organic matter (POM), microbial biomass carbon (MBC), potentially mineralisable nitrogen (PMN), macroaggregate stability, electrical conductivity (EC), sodium adsorption ratio (SAR), pH, inorganic N, P, potassium (K) and magnesium (Mg), bulk density

(BD) and topsoil depth. Water-filled pore space and soil enzyme activity, specifically β-glucosidase activity for its involvement in plant residue degradation, are also added to the recommended list of important soil quality indicators because of their association with soil biological properties and processes (Stott et al. 2010). Tillage factors can also indirectly control residue decomposition processes, because those can influence soil aeration, water content, soil temperature and especially soil aggregate properties. Doran and Parkin have identified good soil quality to accept, hold and release water to plants, streams and groundwater along with other five good characteristics of soil (i) to accept, hold and release nutrients and other chemical constituents, (ii) to promote and sustain root growth, (iii) to maintain suitable soil biotic habitat, (iv) to respond to management and (v) to resist degradation (Doran and Parkin 1994). Those soil processes can moderate water chemistry, atmospheric composition and the bioavailability of elements in soil. Soil properties (e.g., texture, aeration, available moisture, etc.) along with soil fertility can sustain agricultural production, which is dependent on biomass, metabolites and soil microbial activities in the presence of congenial soil water (or moisture) condition (Lehman et al. 2015).

3.4 Soil Microbial Biomass—A Tool for Assessment of Soil Health in the Context of Climate Change

According to Doran and colleagues, soil microbial biomass 'includes both primary and secondary decomposers, aerobic, anaerobic and switch-hitting digestors, highly specialised consumers of gourmet delicacies and feeding trough generalists, finicky occupants of outlandish environmental niches and highly adaptable opportunists, hard-driven frenzied achievers and slow-metabolising plodders, diners of rich, fatty substrates and those eking out an existence gnawing on tough lignaceous scrap. As a group, the community of microbial populations acts without regard for the future, but instead responds quickly to favourable conditions, reproducing and consuming with wild abandon until substrate limitations cause population declines, victims of their collective gluttony. They are in turn cannibalised by their surviving compatriots. The result is a continuous cyclic ballet of nutrient uptake and release that enables less ephemeral life forms in the soil to be supplied with their nutritional needs in a somewhat regulated way' (Doran et al. 1996; Stenberg 1999).

Soil can be defined as a multicomponent and multifunctional complex system. It has definable operating limits and a characteristic spatial configuration. Recognisable soil types originate within a continuum of possibilities, depending on variations in factors of soil formation, viz. parent material, climate and topography based on which dominant soil physical and chemical properties are recognisable. Those are often altered by agricultural interventions of drainage, irrigation, use of lime (for ameliorating soil reaction) and additions of plant nutrients. Soil microbes modify soil structure at submillimetre level by aggregating both mineral and organic constituents through production of extracellular compounds with adhesive properties. Bacteria and fungi produce such soil aggregating adhesive compounds by their feeding mechanisms. Those compounds are useful to aid colony coalescence, to act as protective coatings against drying of soil and to adhere to surfaces of other particles necessary for soil aggregation. Through those soil processes, the changed soil structure and topology of the pore network then work as microbial habitat and thereby affect the distribution and availability of water, delivery of substrate and gases to the organisms and removal of metabolic products from their vicinity and other phenomena associated with microbial activity. Such microbial function is basically controlled by availability of fixed carbon, considered as the major 'currency' of the soil system. That fixed carbon can be manipulated through change in agronomic factors such as crop type, residue and other organic wastes. Organic matter can modulate other soil functions through the provision of surface charges, expressed as the cation exchange capacity, or regulating hydrological properties, viz. wettability. Soil chemical (and physical) properties define the soil as unique system where operate the soil biotic assemblages, adaptive to changes in environmental circumstances, driven by processes of natural selection and are beyond the domain of abiotic systems of the soil (Fierer et al. 2021; Kibblewhite et al. 2008).

Soil biota involve complex adaptive systems by integrating key soil processes for which biological indicators are regarded as an integral component in soil health assessment. On reviewing published information, Ritz and coworkers have noted that an almost exponential increase in potential biological indicators is recognised since the 1970s (Ritz et al. 2009). Environmental functions like microbial, invertebrate or ecological processes are encompassed in many of those biological indicators. Comprehensive elucidation of those considered factors has revealed that synthesising a collective set of biological indicators is limited. SOM and its constituents are generally sorted out as key biological indicators in most of the methods of soil health assessments, because SOM losses are mostly governed by accessibility and availability of SOM to microorganisms than the rate of modification by climate factors (i.e., temperature). From the review, it is revealed that some of the biological factors can be considered as soil biological indicators for soil health assessment like soil C, respiration and soil microbial biomass, microbial and metabolic quotients, enzyme activity, etc. (Allen et al. 2011; Fierer et al. 2021).

3.5   Genetic and Functional Biodiversity of Soils, Soil Health and Climate Change

Present climate change has immense effect on influence of biodiversity on ecosystem functioning of community structure and function of soil biota, particularly the loss of microbial functional groups (Hunt and Wall 2002; Kibblewhite et al. 2008). The earlier studies on potential effects of climate change were primarily focused on assessment of aboveground systems on plant biodiversity and biochemistry (Hunt and Wall 2002), whereas the later studies have found great effect of climate change (such as warming) on the rhizosphere, soil heterotrophic community structure and soil processes, including soil respiration, N mineralisation and ecosystem C functioning (Bardgett et al. 2008; Briones et al. 2009). Possible evolutionary changes are regarded for causing genesis and spread of virulence factors and genes, on adapting to changing climatic factors for environmental survivability and expansion (French et al. 2009). With the expansion of knowledge on microbial community and the drivers of changes in microbial diversity and community structure, aided by the advancements in molecular techniques (Arias et al. 2005; Prosser 2002), it is likely that the genetic and functional biodiversity indicators, particularly concerning food web and nutrient cycling, will be added in the list of soil biological health indicators for improvement in clarifications of climate change impacts on soil health (Allen et al. 2011; Fierer et al. 2021).

3.6 Soil Health Key Indicators for in Situ Soil Health Assessment Under Climate Change

With regard to assessment of soil health under sustainable management practices and predicted climate change scenarios, it is necessary to prioritise among numerous soil health indicators. Recent reviews have focused upon measurement of individual indicators or a suite of indicators to assess changes in soil physical, chemical and/or biological health (Moebius et al. 2007; Zagal et al. 2009), although Stockdale and Watson have noted that such work is necessary to formulate management decisions, rather than simply to compare with the *status quo* based on existing indicator frameworks (Stockdale and Watson 2009).

The interlinked drivers of global climate change are land use change, elevated temperatures, elevated atmospheric CO_2 concentration, increasing atmospheric N deposition, variability in the amount, intensity of rainfall and its distribution, occurrences of extreme climatic conditions and their overall interactions. Like those drivers of global climate change, indicators of soil health are interlinked (Fig. 3.1). Through connecting interactions between climate change, land management and soil health indicators, it has been further revealed that interconnections of many of those drivers (Fig. 3.1) are acting to influence soil health, through their influence on plant primary productivity (PPP), microbial and faunal biomass diversity and activity, including their intracellular and extracellular products like enzymes, as well as interconnections between overall C and N cycles (Bardgett et al. 2008; Das and Varma 2010; Kibblewhite et al. 2008; Wixon and Balser 2009).

Consequently, from most of the studies it is commonly agreed that a minimum data set for soil health assessment should include key indicators that are (1) integrated to soil physical, chemical and biological properties, (2) importantly relatable to soil, (3) applicable to field conditions, (4) sensitive to changes due to land use management and climate variations and (5) accessible to many users (Karlen et al. 2003; Idowu et al. 2009; Lagomarsino et al. 2009; Pattison et al. 2008; Rao and Siddaramappa 2008; Schindelbeck et al. 2008). Above, all the selection of key soil health indicators under climate change scenarios (Fig. 3.1) should reflect the mitigation and adaptive capability of soil and its short- to long-term resilience to climate change.

3.7 Concluding Remarks

Key soil health indicators for assessing soil health affected by climate change and land use management include aggregate stability, SOM, carbon and nitrogen cycling, microbial biomass and activity and microbial fauna and floral diversity. Use of minimum data set of measurable soil health attributes or values, i.e., soil health indicators, makes it possible in understanding and monitoring impacts of climate change on soil health, relating to connections of soil physical, chemical and biological properties to ecological functions in the context of sustainable land management and climate change including the cost of measurement as well as sensitivity of the soil system to changes in climatic factors and land use management and resilience of the system. Such soil health 'tests' are useful for agricultural management and can be promoted within research and government organisations (Nelson et al. 2009; Pattison et al. 2008) for monitoring efforts and policy development (Ritz et al. 2009; Schindelbeck et al. 2008). Generally, soil quality evaluations at the farm, watershed, county, state, regional or national scales are more general and less precise than those made at the point or plot scale (Coyne et al. 2022; Creamer et al. 2022; Karlen et al. 1998; Lehman et al. 2015). So, greater efforts are required to explore individual and interactive effects of drivers of global change under controlled environment and long-term research experiments to undertake climate-adaptive strategies (Allen et al. 2011; Costantini and Mocali 2022; Hassan et al. 2022; Huang et al. 2022) involving suitable soil and water conservation measures on farm land, especially on individual plot for assuring sustainable crop cultivation and food security.

References

Al-Kaisi MM, Lal R (2017) Conservation agriculture systems to mitigate climate variability effects on soil health. In: Al-Kaisi MM, Lowery B (eds) Soil health and intensification of agroecosystems. Elsevier Academic Press, London, pp 79–107

Allen DE, Singh BP, Dalal RC (2011) Soil health indicators under climate change: a review of current knowledge. In: Singh BP, Cowie AL, Chan KY (eds) Soil health and climate change: an overview. Change. Springer Heidelberg, Germany, pp 25–45

Arias ME, González-Pérez JA, González-Vila FJ, Ball AS (2005) Soil health—a new challenge for microbiologists and chemists. Int Microbiol 8:13–21

Bajwa AA, Walsh M, Chauhan BS (2017) Weed management using crop competition in Australia. Crop Prot 95:8–13

Barber SA (1996) Soil nutrient bioavailability: a mechanistic approach. Wiley, New York

Bardgett RD, Freeman C, Ostle NJ (2008) Microbial contributions to climate change through carbon cycle feedbacks. ISME J 2:805–814

Basher LR, Ross CW (2002) Soil erosion rates under intensive vegetable production on clay loam, strongly structured soils at Pukekohe, New Zealand. Aust J Soil Res 40(6):947–961

Blevins RL, Smith MS, Thomas GW, Frye WW (1983) Influence of conservation tillage on soil properties. J Soil Water Conserv 38(3):301–305

Bolan NS, Hedley MJ, White RE (1991) Processes of soil acidification during nitrogen cycling with emphasis on legume based pastures. Plant Soil 134(1):53–63

Borrelli P, Robinson DA, Panagos P, Lugato E, Yang JE, Alewell C, Wuepper D, Montanarella L, Ballabio C (2020) Land use and climate change impacts on global soil erosion by water (2015–2070). Proc Natl Acad Sci 117(36):21994–22001

Brauman KA, Daily GC, Duarte TKE, Mooney HA (2007) The nature and value of ecosystem services: an overview highlighting hydrologic services. Annu Rev Environ Resour 32:67–98

Briones MJI, Ostle NJ, McNamara NP, Poskitt J (2009) Functional shifts of grassland soil communities in response to soil warming. Soil Biol Biochem 41:315–322

Brouder SM, Volenec JJ (2008) Impact of climate change on crop nutrient and water use efficiencies. Physiol Plant 133:705–724

Brussaard L, de Ruiter PC, Brown GG (2007) Soil biodiversity for agricultural sustainability. Agric Ecosyst Environ 121(3):233–244

Charman PEV, Murphy BW (2007) Soils: their properties and management, 3rd edn. Oxford University Press, South Melbourne

Ching PC, Barber SA (1979) Evaluation of temperature effects on K uptake by corn. Agron J 71(6):1040–1044

Christensen NL, Bartuska AM, Brown JH, Carpenter S, D'Antonio C, Francis R, Franklin JF, MacMahon JA, Noss RF, Parsons DJ, Peterson CH, Turner MG, Woodmansee RG (1996) The report of the ecological society of America committee on the scientific basis for ecosystem management. Ecol Appl 6(3):665–691

Costantini EAC, Mocali S (2022) Soil health, soil genetic horizons and biodiversity. J Plant Nutr Soil Sci 185:24–34

Coyne MS, Pena-Yewtukhiw EM, Grove JH, Sant'Anna AC, Mata-Padrino D (2022) Soil health—it's not all biology. Soil Secur 6:100051

Creamer RE, Barel JM, Bongiorno G, Zwetsloot MJ (2022) The life of soils: integrating the who and how of multifunctionality. Soil Biol Biochem 166:108561

Daily GC, Matson PA, Vitousek PM (1997) Ecosystem services supplied by soil. In: Daily GC (ed) Nature's services societal dependence on natural ecosystems. Island Press, Washington, DC, pp 365–374

Dalal RC, Moloney D (2000) Sustainability indicators of soil health and biodiversity. In: Hale P, Petrie A, Moloney D, Sattler P (eds) Management for sustainable ecosystems. Centre for Conservation Biology, Brisbane, pp 101–108

Dalal RC, Lawrence P, Walker J, Shaw RJ, Lawrence G, Yule D, Doughton JA, Bourne A, Duivenvoorden L, Choy S, Moloney D, Turner L, King C, Dale A (1999) A framework to monitor sustainability in the grains industry. Aust J Exp Agric 39:605–620

Dalal RC, Eberhard R, Grantham T, Mayer DG (2003a) Application of sustainability indicators, soil organic matter and electrical conductivity, to resource management in the northern grains region. Aust J Exp Agric 43:253–259

Dalal RC, Wang WJ, Robertson GP, Parton WJ (2003b) Nitrous oxide emission from Australian agricultural lands and mitigation options: a review. Aust J Soil Res 41:165–195

Das SK, Varma A (2010) Role of enzymes in maintaining soil health. In: Shukla G, Varma A (eds) Soil enzymology. Soil biology, vol 22. Springer, Berlin, Heidelberg. https://doi.org/10.1007/978-3-642-14225-3_2

Datiri BC, Lowery B (1991) Effects of conservation tillage on hydraulic properties of a Griswold silt loam soil. Soil Tillage Res 21(3–4):257–271

DEFRA (2005) Impact of climate change on soil functions. Final project report. Research and development. Department for Environment, Food & Rural Affairs (DEFRA), London

Dominati E, Patterson M, Mackay A (2010) A framework for classifying and quantifying the natural capital and ecosystem services of soils. Ecol Econ 69(9):1858–1868

Doran JW (2002) Soil health and global sustainability: translating science into practice. Agric Ecosyst Environ 88:119–127

Doran JW, Parkin TB (1996) Quantitative indicators of soil quality: a minimum data set. In: Doran JW, Jones AD (eds) Methods for assessing soil quality. Soil Science Society of America, Madison, pp 25–37

Doran JW, Sarrantonio M, Liebig MA (1996) Soil Health Sustain Adv Agron 56:1–54

Doran JW, Parkin TB (1994) Defining and assessing soil quality. In: Doran JW, Coleman DC, Bezdicek DF, Stewart BA (eds) Defining soil quality for a sustainable environment vol 35. American Society of Agronomy Special Publication, Madison, pp 3–21. https://doi.org/10.2136/sssaspecpub35.c1

Doran JW, Zeiss MR (2000) Soil health and sustainability: managing the biotic component of soil quality. Agronomy and horticulture—faculty publications 15. Appl Soil Ecol 15:3–11. Available https://digitalcommons.unl.edu/agronomyfacpub/15. Accessed 20 June 2022

Ehlers W, Claupein W (1994) Approaches toward conservation tillage in Germany. In: Carter MR (ed.) Conservation tillage in temperate agroecosystems. Lewis Publishers, Boca Raton, pp 141–165

FAO and ITPS (2015) Status of world's soil resources (SWSR)—main report. Intergovernmental technical panel on soils (ITPS), and FAO, Rome. Available https://www.fao.org/policy-support/tools-and-publications/resources-details/en/c/435200/. Accessed 16 June 2022

FAO (2016) FAO statistical year book 2015. Food and agriculture organization of the United Nations (FAO), Rome

Fierer N, Wood SA, Bueno de Mesquita BP (2021) How microbes can, and cannot, be used to assess soil health. Soil Biol Biochem 153:108111. https://doi.org/10.1016/j.soilbio.2020.108111

French S, Levy-Booth D, Samarajeewa A, Shannon KE, Smith J, Trevors JT (2009) Elevated temperatures and carbon dioxide concentrations: effects on selected microbial activities in temperate agricultural soils. World J Microbiol Biotechnol 25:1887–1900

GISTEMP Team (2018) GISS surface temperature analysis (GISTEMP). NASA goddard institute for space studies, Washington, DC. Available https://data.giss.nasa.gov/gistemp/graphs_v3/Fig.A2.txt. Accessed 20 June 2022

Głab T, Kulig B (2008) Effect of mulch and tillage system on soil porosity under wheat (*Triticumaestivum*). Soil Tillage Res 99(2):169–178

Gonzales-Prieto SJ, Villar MC, Carballas M, Carballas T (1992) Nitrogen mineralization and its controlling factors in various kinds of temperate humid-zone soils. Plant Soil 144:31–44

Hassan W, Li Y, Saba T, Jabbi F, Wang B, Cai A, Wu J (2022) Improved and sustainable agroecosystem, food security and environmental resilience through zero tillage with emphasis on soils of temperate and subtropical climate regions: a review. Int Soil Water Conserv Res, Press. https://doi.org/10.1016/j.iswcr.2022.01.005

Herrick JE, Brown JR, Tugel AJ, Dhaver PL, Havstad KM (1999) Application of soil quality to monitoring and management: paradigms form rangeland ecology. Agro J 94:3–11

Huang B, Yuan Z, Zheng M, Liao Y, Nguyen KL, Nguyen TH, Sombatpanit S, Li D (2022) Soil and water conservation techniques in tropical and subtropical Asia: a review. Sustain 14:5035. https://doi.org/10.3390/su14095035

Hunt HW, Wall DH (2002) Modelling the effects of loss of soil biodiversity on ecosystem function. Glob Change Biol 8:33–50

Idowu OJ, van Es HM, Abawi GS, Wolfe DW, Schindelbeck RR, Moebius-Clune BN, Gugino BK (2009) Use of an integrative soil health test for evaluation of soil management impacts. Renew Agric Food Syst 24:214–224

IPCC (2012) Managing the risks of extreme events and disasters to advance climate change adaptation. In: Stocker TF, Field B, Barros V (eds) A special report of working groups I and II of the intergovernmental panel on climate change. Intergovernmental Panel on Climate Change (IPCC) and Cambridge University Press, Cambridge

Karlen DL, Mausbach MJ, Doran JW, Cline RG, Harris RF, Schuman GE (1997) Soil quality: a concept, definition, and framework for evaluation. Soil Sci Soc Am J 61:4–10

Karlen DL, Gardner JC, Rosek MJA (1998) soil quality framework for evaluating the impact of CRP. J Prod Agric 11:56–60

Karlen DL, Ditzer CA, Andrews SS (2003) Soil quality: why and how? Geoderma 114:145–156

Keesstra SD, Geissen V, Mosse K, Piiranen S, Scudiero E, Leistra M, van Schaik L (2012) Soil as a filter for groundwater quality. Curr Op Environ Sustain 4(5):507–516

Kibblewhite MG, Ritz K, Swift MJ (2008) Soil health in agricultural systems. Philos Transact R Soc 363:685–701

Kinyangi J (2007) Soil health and soil quality: a review. Draft publication. Available https://eco mmons.cornell.edu/bitstream/handle/1813/66582/Soil_Health_Review.pdf. Accessed 20 June 2022

Kladivko EJ (2001) Tillage systems and soil ecology. Soil Tillage Res 61:61–76

Kole C, Muthamilarasan M, Henry R, Edwards D, Sharma R, Abberton M, Batley J, Bentley A, Blakeney M, Bryant J, Cai H, Cakir M, Cseke LJ, Cockram J, de Oliveira AC, De Pace C, Dempewolf H, Ellison S, Gepts P, Greenland A, Hall A, Hori K, Hughes S, Humphreys MW, Iorizzo M, Ismail AM, Marshall A, Mayes S, Nguyen HT, Ogbonnaya FC, Ortiz R, Paterson AH, Simon PW, Tohme J, Tuberosa R, Valliyodan B, Varshney RK, Wullschleger SD, Yano M, Prasad M (2015) Application of genomics-assisted breeding for generation of climate resilient crops: Progress and prospects. Front Plant Sci 6:563. https://doi.org/10.3389/fpls.2015.00563

Lagomarsino A, Moscatelli C, Tizio AD, Mancinelli R, Grego S, Marinari S (2009) Soil biological indicators as a tool to assess the short-term impact of agricultural management on changes in organic C in a Mediterranean environment. Ecol Indic 9:518–527

Lal R (2000) Mulching effects on soil physical quality of an alfisol in western Nigeria. Land Degrad Develop 11(4):383–392

Lal R (2003) Soil erosion and the global carbon budget. Environ Int 29(4):437–450

Lal R (2004) Soil carbon sequestration impacts on global climate change and food security. Sci 304(5677):1623–1627

Lal R (ed.) (1999) Soil quality and soil erosion. Soil water conservation society, Ankeny

Lal R (2011) Soil health and climate change: an overview. In: Singh BP, Cowie AL, Chan KY (ed) Soil health and climate change: an overview. Change. Springer Heidelberg, pp 3–24

Lavelle P, Decaëns T, Aubert M, Barot S, Blouin M, Bureau F, Margerie P, Mora P, Rossi JP (2006) Soil invertebrates and ecosystem services. Eur J Soil Biol 42:S3–S15

Lehman RM, Cambardella CA, Stott DE, Acosta-Martinez V, Manter DK, Buyer JS, Maul JE, Smith JL, Collins HP, Halvorson JJ, Kremer RJ, Lundgren JG, Ducey TF, Jin VL, Karlen DL (2015) Understanding and enhancing soil biological health: the solution for reversing soil degradation. Sustain 7:988–1027. https://doi.org/10.3390/su7010988

Lorenz K, Lal R (2014) Soil organic carbon sequestration in agroforestry systems. A review. Agron Sustain Develop 34(2):443–454

Luo Z, Feng W, Luo Y, Baldock J, Wang E (2017) Soil organic carbon dynamics jointly controlled by climate, carbon inputs, soil properties and soil carbon fractions. Glob Change Biol 23(10):4430–4439

Mackay AD, Barber SA (1984) Soil temperature effects on root growth and phosphorus uptake by corn. Soil Sci Soc Am J 48(4):818–823

Mackay AD, Barber SA (1985) Soil moisture effect on potassium uptake by corn. Agron J 77(4):524–527

Magdoff F (2001) Concept, components and strategies of soil health in agroecosystems. J Nematol 33:169–172

Mangalassery S, Sjögersten S, Sparkes DL, Sturrock CJ, Craigon J, Mooney SJ (2014) To what extent can zero tillage lead to a reduction in greenhouse gas emissions from temperate soils? Sci Rep 4:4586. https://doi.org/10.1038/srep04586

MEA(2005) Ecosystems and human well-being: synthesis. Mil-lennium ecosystem assessment (MEA), World resources Institute, Washington, DC. Island Press, Washington, DC

Mehra P, Singh BP, Kunhikrishnan A, Cowie AL, Bolan N (2018) Soil health and climate change: a critical nexus. In: Reicosky D (ed) Managing soil health for sustainable agriculture, volume 1: fundamentals. Burleigh Dodds Science Publishing, Cambridge

Merrill SD, Black AL, Bauer A (1996) Conservation tillage affects root growth of dryland spring wheat under drought. Soil Sci Socf Am J 60:575–583

Misra AK (2014) Climate change and challenges of water and food security. Int J Sustain Built Environ 3(1):153–165

Moebius BN, van Es HM, Schindelbeck RR, Idowu OJ, Clune DJ, Thies JE (2007) Evaluation of laboratory-measured soil properties as indicators of soil physical quality. Soil Sci 172:895–912

Montanari A, Young G, Savenije HHG, Hughes D, Wagener T, Ren LL, Koutsoyiannis D, Cudennec C, Toth E, Grimaldi S, Blöschl G, Sivapalan M, Beven K, Gupta H, Hipsey M, Schaefli B, Arheimer B, Boegh E, Schymanski SJ, Di Baldassarre G, Yu B, Hubert P, Huang Y, Schumann A, Post DA, Srinivasan V, Harman C, Thompson S, Rogger M, Viglione A, McMillan H, Characklis G, Pang Z, Belyaev V (2013) Panta Rhei—everything flow: change in hydrology and society—the IAHS scientific decade 2013–2022. Hydrol SciJ 58(6):1256–1275. https://doi.org/10.1080/026 26667.2013.809088

Monteith J, Unsworth M (2013) Principles of environmental. Physics plant, animals and the atmosphere, 4th edn. Academic Press, Boston

Nannipieri P, Ascher J, Ceccherini M, Landi L, Pietramellara G, Renella G (2003) Microbial diversity and soil functions. Eur J Soil Sci 54:655–670

Nelson KL, Lynch DH, Boiteau G (2009) Assessment of changes in soil health throughout organic potato rotation sequences. Agric Ecosyst Environ 131:220–228

Nolin MC, Wang C, Caillier MJ (1989) Fertility grouping of montreal lowlands soil mapping units based on selected soil characteristics of the plow layer. Can J Soil Sci 69:525–541

Nuttall JG (2007) Climate change—identifying the impacts on soil and soil health. Department of Primary Industries, Future Farming Systems Research Division, Victoria

Pattison AB, Moody PW, Badcock KA, Smith LJ, Armour JA, Rasiah V, Cobon JA, Gulino L-M, Mayer R (2008) Development of key soil health indicators for the Australian banana industry. Appl Soil Ecol 40:155–164

Petersen SO, Schjønning P, Thomsen IK, Christensen BT (2008) Nitrous oxide evolution from structurally intact soil as influenced by tillage and soil water content. Soil Biol Biochem 40(4):967–977

Poeplau C, Don A (2015) Carbon sequestration in agricultural soils via cultivation of cover crops—a meta-analysis. Agric Ecosyst Environ 200:33–41

Potter S, Cabbage M, McCarthy L (2017) NASA, NOAA data show 2016 warmest year on record globally. National Aeronautics and Space Administration, Washington, DC. Available https://www.nasa.gov/press-release/nasa-noaa-data-show-2016-warmest-year-on-rec ord-globally. Accessed 20 June 2022

Prosser JI (2002) Molecular and functional diversity in soil micro-organisms. Plant Soil 244:9–17

Qian B, Gregorich EG, Gameda S, Hopkins DW, Wang XL (2011) Observed soil temperature trends associated with climate change in Canada. J Geophys Res 116:D2106. https://doi.org/10.1029/2010JD015012

Rao BKR, Siddaramappa R (2008) Evaluation of soil quality parameters in a tropical paddy soil amended with rice residues and tree litters. Eur J Soil Biol 44:334–340

Rengel Z (2011) Soil pH, soil health and climate change. In: Singh BP, Cowie AL, Chan KY (eds) Soil health and climate change soil biology series 29. Springer, Berlin, pp 69–85

Rhodes CJ (2014) Soil erosion, climate change and global food security: challenges and strategies. Sci Progr 97(2):97–153

Riley J (2001) Multidisciplinary indicators of impact and change: key issues for identification and summary. Agric Ecosyst Environ 87:245–259

Ritz K, Black HIJ, Campbell CD, Harris JA, Wood C (2009) Selecting biological indicators for monitoring soils: a framework for balancing scientific and technical opinion to assist policy development. Ecol Indic 9:1212–1221

Rogers CE, McCarty JP (2000) Climate change and ecosystems of the mid-Atlantic region. Clim Res 14(3):235–244

Salem HM, Valero C, Muñoz MÁ, Rodríguez MG, Silva LL (2015) Short-term effects of four tillage practices on soil physical properties, soil water potential, and maize yield. Geoderma 237–238:60–70

Sarker JR, Singh BP, Cowie AL, Fang Y, Collins D, Dougherty WJ, Singh BK (2018a) Carbon and nutrient mineralisation dynamics in aggregate-size classes from different tillage systems after input of canola and wheat residues. Soil Biol Biochem 116:22–38

Sarker JR, Singh BP, Dougherty WJ, Fang Y, Badgery W, Hoyle FC, Dalal RC, Cowie AL (2018b) Impact of agricultural management practices on the nutrient supply potential of soil organic matter under long-term farming systems. Soil Tillage Res 175:71–81

Schindelbeck RR, van Es HM, Abawi GS, Wolfe DW, Whitlow TL, Gugino BK, Idowu OJ, Moebius-Clune BN (2008) Comprehensive assessment of soil quality for landscape and urban management. Landsc Urban Plan 88:73–80

Schjønning P, Elmholt S, Christensen BT (2004) Soil quality management—concepts and terms. In: Schjønning P, Elmholt S, Christensen BT (eds) Managing soil quality: challenges in modern agriculture. CAB International, Oxon

Singh BP, Setia R, Wiesmeier M, Kunhikrishnan A (2018a) Agricultural management practices and soil organic carbon storage. In: Singh B (ed) Soil carbon storage: modulators, mechanisms and modeling, 1st edn. Academic Press, London, pp 207–244

Singh BP, Cowie AL, Chan KY (2011) Soil health and climate change. Soil biology series 29. Springer, Berlin

Sojka RE, Bjorneberg DL, Trout TJ, Strelkoff TS, Nearing MA (2007) The importance and challenge of modelling irrigation-induced erosion. J Soil Water Conserv 62(3):153–162

Stenberg B (1999) Monitoring soil quality of arable land: microbiological indicators. Acta Agric Scandinavica Sect B Soil Plant Sci 49:1–24

Stenberg B, Pell M, Torstensson L (1998) Integrated evaluation of variation in biological, chemical and physical soil properties. Ambio 27:9–15

Stockdale EA, Watson CA (2009) Biological indicators of soil quality in organic farming systems. Renew Agric. Food Syst 24:308–318

Stott DE, Andrews SS, Liebig MA, Wienhold BJ, Karlen DL (2010) Evaluation of β-glucosidase activity as a soil quality indicator for the soil management assessment framework (SMAF). Soil Sci Soc Am J 74:107–119

Thornton PK, Ericksen PJ, Herrero M, Challinor AJ (2014) Climate variability and vulnerability to climate change: a review. Glob Change Biol 20(11):3313–3328

Titeux N, Henle K, Jean-Baptiste M, Regos A, Geijzendorffer IR, Cramer W, Verburg PH, Brotons L (2016) Biodiversity scenarios neglect future land-use changes. Glob Change Biol 22(7):2505–2515

Triplett GB, Dick WA (2008) No-tillage crop production: a revolution in agriculture! Agron J100:S153–S165

Union E (2014) Study on soil and water in a changing environment. Final Report, Chap 1:33–46. https://doi.org/10.2779/20608

Ussiri DAN, Lal R, Jarecki MK (2009) Nitrous oxide and methane emissions from longterm tillage under a continuous corn cropping system in Ohio. Soil Tillage Res 104(2):247–255

Wienhold BJ, Karlen DL, Andrews SS, Stott DE (2009) Protocol for soil management assessment framework (SMAF) soil indicator scoring curve development. Renew Agric Food Syst 24:260–266

Wild A (1993) Soils in relation to the environment. In: Wild A (ed) Soils and the environment. Cambridge University Press, Cambridge, pp 107–108

Wixon DL, Balser TC (2009) Complexity, climate change and soil carbon: a systems approach to
   microbial temperature response. Syst Res Behav Sci 26:601–620
Zagal E, Muñoz C, Quiroz M, Córdova C (2009) Sensitivity of early indicators for evaluating quality
   changes in soil organic matter. Geoderma 151:191–198

Chapter 4
Effect of Soil on Water Quality

Abstract More soil losses occur from open soil surface than the covered surface area. Depending on local societal needs, vegetative cover may be followed for reducing soil erosion from crop fields. All agricultural and forest activities including development works like road construction require careful operations for minimising erosion and sedimentation in streams and managing surface and groundwater qualities.

Keywords Aquatic ecosystem · Forest · Land use · Soil · Topography · Vegetation · Water conservation

Abbreviations

BOD	Biological oxygen demand
BGA	Blue green algae
CAST	Council for Agricultural Science and Technology
DO	Dissolved oxygen
FAO	Food and Agriculture Organization of the United Nations
HEPP	Hydroelectric power plant
K	Potassium
LULC	Land use land cover
P	Phosphorus
SOM	Soil organic matter
SWAT	Soil and Water Assessment Tool
UNESCO	United Nations Educational, Scientific and Cultural Organization
Zn	Zinc

© The Author(s), under exclusive license to Springer Nature Switzerland AG 2022
S. Panda, *Soil and Water Conservation for Sustainable Food Production*,
Chemistry of Foods, https://doi.org/10.1007/978-3-031-15405-8_4

4.1 Introduction

Water is another name of life, as they say, water is life. In other words, survivability of life means availability of water. Fresh water determines sustenance of life on the earth. Though 70% of earth's surface is covered with water, only a small part of it about 1.7% out of total water is in poles and only about 0.77% of total global water is in the form of fresh groundwater, lakes, rivers and soil moisture (Baker et al. 2016; Chang 2013; Shiklomanov 1998; Subramanya 2008; UNESCO 1971). Mankind depends on such a scarce natural resource for daily life and agricultural water use for which quality of water is the prime concern. Agricultural land use, on the other hand, affects quality of water for which soil has a major role. Quality and quantity of water available in a watershed are influenced by (1) land use status, e.g., settlements, agriculture, pasture, forest, etc., (2) vegetative cover, e.g., trees, grasses, bushes, agroforestry, etc., (3) climate, (4) geology, (5) topography and (6) soil physical, chemical and biological properties. Thus, those factors affect flow and quality of water in a stream (Ballantine et al. 2009; Flood et al. 2022; Hawthorne et al. 2013; Johnson et al. 1997; Koralay et al. 2018; Rehman et al. 2015; Sensoy and Kara 2013; Sthiannopkao et al. 2006). Water quality has a great binding on aquatic ecosystem. The physical, chemical and biological quality of water depends on precipitation, soil erosion and sources of eroded soil from various types of land uses. In that way, water quality has great impact on temperature, dissolved oxygen, nutrients and sediment loads in water and thereby aquatic ecosystem and watershed hydrology in general (Issaka and Ashraf 2017; Osterholz et al. 2021; Sthiannopkao et al. 2006, Welde and Gebremariam 2017; Zimnicki et al. 2020). In soil water quality index by Doran and Parkin, both surface and groundwater qualities are included. Soil management practices including uses of pesticides and fertilisers, on the other hand, can also influence atmospheric quality either by producing or consuming atmospheric gases like carbon dioxide, methane, nitrous oxide, etc. (Doran and Parkin 1994). It is revealed that increase in soil organic carbon content has least effect on soil water retention and such increment is larger in sandy soils followed by loams and is least in clay soils (CAST 1992; Doran and Safley 1997; Minasny and Mcbratney 2017; Rolston et al. 1993).

4.2 Effect of Geology on Water Quality

Type and structure of the rocks that make up the land structure in a basin influence the soil type formed in the basin. The amount of soil erosion from the field is ascertained based on the type of soil. Erosion of bedrock is influenced by the type of soil overlying it, for example, more erosion due to overlying sandy soil than loamy soil. Geological structures also cause differences in total amount of nutrients like phosphorus coming out with sediments from different land situations, though those are also influenced by climatic variations and land use types (Ballantine et al. 2009). Chemical composition and density of the sediment are the result of rock types, clay minerals, soil structure, forest destruction and agricultural practices followed in the watershed (Duarte and Gioda 2014).

4.3   Effect of Topography on Water Quality

Soil erosion is highly influenced by topographic features of which slope and length of the land are most important. Those two features are dominant factors for causing surface runoff and soil erosion, and their rates vary with rainfall variation and land characteristics (Rehman et al. 2015; Sensoy and Kara 2013).

4.4   Effect of Soil Erosion and Water Quality as Influenced by Climate

Intense rainfalls or fast-moving winds cause soil erosion from fallow land without vegetation cover. Irregularity in climatic pattern also adds to the cause of soil erosion process, though naturally it varies seasonally. The amounts of suspended soils remain higher in wet periods than in dry periods. From mineralogical and chemical analysis of suspended solids in water, those are found to contain elements like potassium (K), phosphorus (P) and zinc (Zn) as their constituents and within which P remains to be higher in wet periods (Duarte and Gioda 2014).

4.5   Effect of Soil Properties on Water Quality

Physical, chemical and biological soil properties such as texture, structure, porosity, depth, amount of organic matter, soil salinity and vegetation cover are deciding factors in causing soil to suffer erosion. Soil organic matter (SOM) is one of the major deciding factors for susceptibility of soil to erosion. Amount of SOM is inversely related with amount of soil erosion from a land. In contrast with that, SOM improves water holding capacity, aggregation and cementing properties including other soil physical properties like structure, permeability and porosity. Thus, more water can enter into soil profile, and after soaking into soil, excess water can reach stream safely (Chang 2013; Koralay et al 2018).

4.6   Effect of Soil Erosion on Water Quality as Influenced by Vegetation Cover

Vegetation cover has great influence on soil erosion process. Plant cover of trees, shrubs and herbs can form a mechanical barrier against rainfall and runoff by reducing the effects of climate, topography and soil itself. It can increase infiltration and water storage, decrease soil erosion and maintain congenial condition in soil moisture storage (Fen-Li 2006; García-Ruiz 2010; Zhang et al. 2015). Thus, forest floor can save soil from erosion loss by reducing impact of rainfall, favouring soil and ground-water storages and controlling erosivity of overland flow of water and its draining into the stream (Biddoccu et al 2020; Guerra et al. 2020; Miyata et al. 2009).

4.7   Effect of Watershed on Water Quality

A watershed is a topographic structure, separated from adjacent ones by a continuous ridgeline, and it drains out excess rainfall water through a single outlet. Watershed management is targeted towards control of erosion through management of flood and other undesirable rainfall events in a watershed, aimed at producing water of highest quality and quantity on considering socio-economic conditions and managing natural resources in the watershed. Various types of land use land cover (LULC) patterns like forest, agriculture, pasture, etc., characterise watershed hydrology in general. Those land use patterns greatly influence soil erosion and water quality of that watershed. Consequently, soil erosion and water quality are significantly correlated with topographical and hydrological characteristics of the watershed, including various facilities on the stream such as hydroelectric power plants (HEPP) and land use patterns. Soil characteristics, water quality and quantity are badly affected especially in the areas the of farm, domestic (rural or urban), converted areas from forest and pasture lands and also in areas developed under social set ups like highway and runway constructions, HEPP, industrialisation, domestic waste processing, etc. Such land conversions and social sector-related constructions may reduce the resistance of soil to external factors and thereby increase its tendency to suffer from erosion. Such soil processes may significantly increase sedimentation in streams and negatively affect their water quality which is important for all living beings both in terrestrial and aquatic ecosystems. For that reason, sedimentation control should be given emphasis in any basin planning (Issaka and Ashraf 2017; Koralay et al. 2018).

4.7.1 Effect of Soil Erosion on Water Quality of Aquatic Ecosystem and Watershed Hydrology

Soil erosion can affect the hydrological cycle in a watershed by changing soil compaction, infiltration, water holding capacity, vegetation pattern and evapotranspiration. Minerals and oxygen are fed into underground and surface waters through water flow in rivers and streams. Dissolved oxygen and mineral nutrients, released from the chemical and biological processes in presence of water, are important for sustenance of life continuity in terrestrial and aquatic ecosystems and soil environments.

Increased soil erosion is the cause of increase in sediment and/ or suspended solids in the stream which lead to so many difficulties in the environment of the stream and ultimate reduction in the density of fish and invertebrates. This is also true for wetlands and other inland aquatic environments. The favourable levels of suspended solids for fish should be around 25 mg L^{-1}, and the unfavourable condition rises above 100 mg L^{-1}. Not only the concentration but also both the flux and concentration of suspended solids are important for hatching of eggs and fish survivability (Fig. 4.1). Deposition of suspended solids on the hatching stones and pebbles fills up the habitats, i.e., hatching spaces and protective holes for different growing stages of fish and other benthic and amphibian fauna right from eggs to fingerling adults including stages of larva, fry or minnow and juvenile. Gills of fish and other breathing organs of those wetland and aquatic fauna are choked due to deposition of high amounts of suspended solids and thereby making it difficult for them to survive there or to leave the affected habitats. The light input required for photosynthesis by aquatic plants may be reduced with increase in turbidity caused due to presence of more suspended solids. This can lead to a reduction in the amount of nutrients needed by aquatic organisms. Increased concentration of suspended solids causes increase in turbidity, and thereby, transparency to light as well as light intensity in water is reduced. That condition prevents aquatic green flora from potentially yielding primary production through photosynthesis that affects normal food supply for aquatic fauna and fishes. Poisoning of benthic organisms is caused due to inflow of high amounts of suspended solids from streams and especially from irrigation channels. Along with those difficulties, accumulation of suspended solids in streams, rivers, wetlands and other aquatic habitats also causes deprivation in oxygen levels there. All those factors affect survivability of fishes and all other benthic and other aquatic fauna and flora and the overall environment as well (Belal et al. 2016; Bregnballe 2015; Chang 2013; Pfeifer et al. 2000; Reynolds et al. 1989; Townsend et al. 2009).

The use of cover crops in agricultural lands is widely considered as an effective soil and water conservation measure. Because cover crops can afford various supporting and provisioning ecosystem services of reduction of runoff and erosion processes, increasing soil organic matter, improvement of field trafficability, water purification, water supply, biodiversity conservation including weed control, pest and disease regulation (Garcia et al. 2018; Hall et al 2020; Winter et al 2018). From several

Fig. 4.1 Riverbed
sedimentation

studies, it is revealed that large reductions in erosion rates are found by raising a
cover crop in agricultural lands (Fig. 4.2). It has been demonstrated after a 14-year
monitoring experiment that the reduction in average soil losses by one order of
magnitude (from 13.85 t ha^{-1} year^{-1}to 1.8 t ha^{-1} year^{-1}) from large runoff plots
under vineyards with permanent grass cover in Italy. Similar results on reduction of
58% in soil losses (from an average of 48.1 t ha^{-1} year^{-1}1 on tilled plots to 27.9 t
ha^{-1} year^{-1}) from 3-year long experiment on large runoff plots under vineyards with
cover crop in Portugal and France (Biddoccu et al. 2016; Gomez et al. 2011).

Sometimes cover crop treatment may fail due to extreme rainfall events for which
large soil losses may occur (Gomez et al. 2011). Local societal needs may also
be the deciding factor for practising good protection by cover crop throughout the
rainy season; e.g., for successfully growing target crop (like grape vine). Necessity for permanent cover crop in Mediterranean dry region is not followed for
avoiding competition for soil water between those two crops (Celette et al. 2008;
Ruiz-Colmenero et al. 2011) for which farmers in that region are used to establish
bare soil vineyards or temporary cover crops of 4 to 7 months each year (Salome
et al. 2016). In contrast to that situation, in more humid conditions of central Europe
permanent cover crops are commonly practised (Biddoccu et al. 2020; Lefèvre et al.
2020).

Fig. 4.2 Cover crop of black
gram

4.7.1.1 Effect of Soil Erosion on Water Temperature

Conversion of agricultural lands into settlements, clearing of the riparian zone, forest and pasture lands into agricultural areas, production activities and construction of road and runways primarily increase soil erosion, thereby affect water quality (Morris et al. 2016; Rice and Wallis 1962) and reduces macro- and microflora and fauna, vegetation in stream and other aquatic and terrestrial living beings. Canopy cover and the forest floor with roots and organic materials can increase water retention, infiltration and permeability, and thus, those can reduce surface runoff and associated soil erosion. Due to decreasing effect of forest floor on the maximum flow in the stream hydrograph, peak point time is delayed; thereby, sediment load is decreased in the stream (Li et al. 2014). Eroded soil, thus lost from the original land settings, is transported in water bodies and streams and increases suspended solid materials in water; thereby, increase in surface water temperature is caused due to increase in suspended solids. Higher water temperature brings in changes in water and soil physicochemical interactions and characteristics, for example, reductions in water leaching and rate of sedimentation of suspended solids. With the rise in water temperature, there will be decrease in concentration of dissolved oxygen and increase in more evaporation of water. Such aquatic environment is congenial for proliferation of blue green algae (BGA) and other destructive microorganisms for which water quality will further degrade, though that BGA biomass may be recycled as mulching materials for soil moisture conservation in the crop fields. Usually, aquatic organisms can adapt to seasonal changes in water temperature, but such nonseasonal increase in water temperature can affect the growth of fish and invertebrates and that increase by one or two degrees negatively affects the lives of fish, i.e., their growth, migration, movement, reproduction, laying of eggs, etc., and after exceeding certain optimal condition in water temperature, fish diseases (like changes in appearance) occur and may turn to be epidemic resulting in numerous fish deaths (Gyawali et al. 2013; Li and Migliaccio 2011; MacDonald et al. 1991).

4.7.1.2 Effect of Soil Erosion on Dissolved Oxygen in Water

Aquatic lives are regulated by concentration of dissolved oxygen (DO) in that environment and that should be > 5 mg L^{-1}. Survival of aquatic insects and fish eggs is reduced in low DO condition. Covering of eggs, larvae and fish habitat by accumulation of sediment also causes deprivation of oxygen in stream, especially in pebbly bedded stony water flow; thereby, metabolism is affected with the release of ammonia and carbon compounds poisonous for aquatic organisms (Chang 2013).

4.8   Effect of Land Use Land Cover on Water Quality

Land use and land cover classification is one of the most important interventions for ascertaining soil erosion. Among different land uses, highest erosion from paddy fields and the least erosion from forest areas are observed, and most important effect on sediment yield is due to the way and the traces made by the machines used in operation of forest works and removal of vegetation and protective cover on the soil due to logging. Unwanted effect on water quality is also caused when forestry works are performed without complying with the required rules on forestry activities (Brown and Binkley 1994). A Soil and Water Assessment Tool (SWAT) model study has revealed that there is direct relation with the increase in naked land and agricultural areas with increase in average annual stream flow and amount of sediment yield (Welde and Gebremariam 2017). Ecological processes in rivers and lakes are affected by nitrogen and phosphorus levels in water. Though those nutrient increase can raise the level of fertility in water and consequent decrease in DO level, thereby causing decrease in species diversity. Very high nutrient (especially nitrates and phosphates) addition causes dangerous spawning of the algae (i.e., eutrophication or algal bloom) and aquatic weeds limiting entry of sunlight, increase in turbidity and biological oxygen demand (BOD), thus decreasing DO level in water. Increase in the amount of phosphorus in the fluvial sediment is reported from the watershed dominated by the rural community and industrial estates (Owens and Walling 2002). As the soil erosion reduces soil fertility, it increases sediment loads with soil, nutrients and agrochemicals, thus decreasing both water and environmental qualities (Brown 1985; Sthiannopkao et al. 2006).

As resistance of the soil to external factors is reduced due to various land conversions, tendency of soil to suffer from erosion increases, and there occurs significant reductions in sediments reaching the streams. As the various land conversion activities increase the amount of sediment in the streams, various land use land cover, climate, geology, topography, physical, chemical and biological soil properties, etc., influence soil erosion in a watershed. For example, increase in algal population, and also increase in sediment and organic matter transported to the mainstream due to the decrease in vegetation cover (Molina and Campo 2012). Due to such situational emergence and consequent deterioration in water quality and decrease in populations of invertebrates, fishes and other aquatic organisms, biodiversity requires smallest intervention of site-specific planning in the watershed with due importance for values of water resources and its quality and reduction in soil erosion. For example, riparian zone can reduce water temperature by the effect of shadow, increase in water quality by decreasing sediment yield and filter of fertiliser and chemicals. Thus, riparian zone, located in narrow areas at the edges of stream, has very significant advantages on water quality and soil erosion. So, such area requires careful working on land for economic uses through watershed-based planning (Gyawali et al. 2013; Johnson et al. 1997; Mello et al. 2017; Sthiannopkao et al. 2007; Wear and Greis 2002).

4.9   Concluding Remarks

Soil erosion is the prime factor of sustained water quality. More soil losses occur from open soil surface than the covered surface area. Vegetative cover may be useful in conserving soil from erosion losses, though it depends on local societal needs. Forest operations like logging, transport, construction of roads, etc., should consider careful works with minimum disturbances in the environment. In agricultural fields, vegetative cover, so far accepted by the farmer, may be practised which will be helpful in conserving loss of soil from individual crop fields. As the eroded soil takes away nutrients and agrochemicals with it, conserving plot-wise soil loss has the cumulative effect on sustaining water quality and healthy environment.

References

Baker B, Aldridge C, Ome A (2016) Water: availability and use. Mississippi state university extension, Publication 3011 (POD-11-16). Mississippi State University, Mississippi State

Ballantine DJ, Walling DE, Collins AL, Leeks GJL (2009) The content and storage of phosphorus in fine grained channel bed sediment in contrasting lowland agricultural catchments in the UK. Geoderma 151:141–149

Belal AA, El-Sawy MA, Dar MA (2016) The effect of water quality on the distribution of macro-benthic fauna in Western Lagoon and Timsah Lake, Egypt. I. Egypt J Aquat Res 42:437–448

Biddoccu M, Ferraris S, Opsi F, Cavallo E (2016) Long-term monitoring of soil management effects on runoff and soil erosion in sloping vineyards in Alto Monferrato (North-West Italy). Soil and Tillage Research 155: 176–189. https://doi.org/10.1016/j.still.2015.07.005

Biddoccu M, Guzman G, Capello G, Thielke T, Strauss P, Winter S, Zaller JG, Nicolai A, Cluzeau D, Popescu D, Bunea C, Hoble A, Cavallo E, Gomez JA (2020) Evaluation of soil erosion risk and identification of soil cover and management factor (C) for RUSLE in European vineyards with different soil management. Int Soil Water Conserv Res 8(4):337–353. https://doi.org/10.1016/j.iswcr.2020.07.003

Bregnballe J (2015) A guide to recirculation aquaculture an introduction to the new environmentally friendly and highly productive closed fish farming systems. The Food and Agriculture Organization of the United Nations (FAO), Rome, and EUROFISH International Organisation, Copenhagen, p 56. Available https://www.fao.org/3/i4626e/i4626e.pdf. Accessed 21 June 2022

Brown TC, Binkley D (1994) Effect of management on water quality in North American forests. General technical report RM–248. Rocky mountain forest and range experiment station, forest service, United States Department of Agriculture, Fort Collins, CO

Brown GW (1985) Controlling nonpoint source pollution from silvicultural operations: what we know and don't know. In: Proceedings of a national conference: perspectives on nonpoint source pollution. Environmental Protection Agency, Washington, DC, pp 332–333

CAST (1992) Preparing US agriculture for global climate change. Task force report no. 119. Council for Agricultural Science and Technology (CAST), Ames

Celette F, Gaudin R, Gary C (2008) Spatial and temporal changes to the water regime of a Mediterranean vineyard due to the adoption of cover cropping. Eur J Agron 29:153–162. https://doi.org/10.1016/j.eja.2008.04.007

Chang M (2013) Forest hydrology: an introduction to water and forests, 3rd edn. CRC Press, Boca Raton, p 72

Doran JW, Parkin TB (1994) Defining and assessing soil quality. In: Doran JW, Coleman DC, Bezdicek DF, Stewart BA (eds) Defining soil quality for a sustainable environment, vol 35.

American Society of Agronomy Special Publication, Madison, pp 3–21. https://doi.org/10.2136/sssaspecpub35.c1

Doran JW, Safley M (1997) Defining and assessing soil health and sustainable productivity. In: Pankhurst C, Doube BM, Gupta V (eds) Biological indicators of soil health. CABI Publishing, Wallingford, pp 1–28

Duarte AF, Gioda A (2014) Inorganic composition of suspended sediments in the Acre River, Amazon Basin, Brazil. Lat Am J Sedimentol Basin Anal 21(1):3–15

Fen-Li Z (2006) Effect of vegetation changes on soil erosion on the loess plateau. Pedosphere 16(4):420–427

Flood MT, Hernandez-Suarez JS, Nejadhashemi AP, Martin SL, Hyndman D, Rose JB (2022) Connecting microbial, nutrient, physiochemical, and land use variables for the evaluation of water quality within mixed use watersheds. Water Res 219:118526. https://doi.org/10.1016/j.watres.2022.118526

Garcia L, Celette F, Gary C, Ripoche A, Valdés-Gómez H, Metay A (2018) Management of service crops for the provision of ecosystem services in vineyards: a review. Agric Ecosyst Environ 251:158–170. https://doi.org/10.1016/j.agee.2017.09.030

García-Ruiz JM (2010) The effects of land uses on soil erosion in Spain: a review. CATENA 81:1–11

Gomez JA, Llewellyn C, Basch G, Sutton PB, Dyson JS, Jones CA (2011) The effects of cover crops and conventional tillage on soil and runoff loss in vineyards and olive groves in several Mediterranean countries. Soil Use Manag 27:502–514. https://doi.org/10.1111/j.1475-2743.2011.00367.x

Guerra CA, Rosa IMD, Valentini E, Wolf F, Filipponi F, Karger DN, Xuan AN, Mathieu J, Lavelle P, Eisenhauer N (2020) Global vulnerability of soil ecosystems to erosion. Landsc Ecol 35:823–842. https://doi.org/10.1007/s10980-020-00984-z

Gyawali S, Techato K, Yuangyai C, Musikavong C (2013) Assessment of relationship between land uses of riparian zone and water quality of river for sustainable development of river basin—a case study of U-Taoao river basin, Thailand. Procedia Environ Sci 17:291–297. https://doi.org/10.1016/j.proenv.2013.02.041

Hall RM, Penke N, Kriechbaum M, Kratschmer S, Jung V, Chollet S, Guernion M, Nicolai A, Burel F, Fertil A, Lora Á, Sánchez-Cuesta R, Guzmán G, Gómez J, Popescu D, Hoble A, Bunea C-I, Zaller JG, Winter S (2020) Vegetation management intensity and landscape diversity alter plant species richness, functional traits and community composition across European vineyards. Agric Syst 177:102706. https://doi.org/10.1016/j.agsy.2019.102706

Hawthorne SND, Lane PNJ, Bren LJ, Sims NC (2013) The long term effects of thinning treatments on vegetation structure and water yield. For Ecol Manag 310:983–993

Issaka S, Ashraf MA (2017) Impact of soil erosion and degradation on water quality: a review. Geol Ecol Landsc 1(1):1–11. https://doi.org/10.1080/24749508.2017.1301053

Johnson LB, Richards C, Host GE, Arthur JW (1997) Landscape influences on water chemistry in Midwestern stream ecosystems. Freshwater Biol 37:193–208

Koralay N, Kara O, Kezik U (2018) Effects of run-of-the-river hydropower plants on the surface water quality in the Solakli stream watershed, Northeastern Turkey. Water Environ J 32:412–421. https://doi.org/10.1111/wej.12338

Lefèvre C, Cruse RM, Cunha dos Anjos LH, Calzolari C, Haregeweyn N (2020) Guest editorial – soil erosion assessment, tools and data: A special issue from the Global symposium on soil Erosion 2019. International Soil and Water Conservation Research 8 (2020) 333–336. https://doi.org/10.1016/j.iswcr.2020.11.004

Li Y, Migliaccio K (2011) Water quality concepts, sampling and analyses. CRC Press, LLC, Boca Raton, Taylor and Francis Group

Li X, Niu J, Xie B (2014) The effect of leaf litter cover on surface runoff and soil erosion in Northern China. PLoS ONE 9(9):e107789. https://doi.org/10.1371/journal.pone.0107789

MacDonald LH, Smart AW, Wissmar RC (1991) Monitoring guidelines to evaluate effects of forestry activities on streams in the Pacific Northwest and Alaska. EPA 910/9–91–001. Environmental Protection Agency, Washington, DC

Mello KD, Randhir TO, Valente RA, Vettorazzi CA (2017) Riparian restoration for protecting water quality in tropical agricultural watersheds. Ecol Eng108, Part B:514–524. https://doi.org/10.1016/j.ecoleng.2017.06.049

Minasny B, Mcbratney AB (2017) Limited effect of organic matter on soil available water capacity. Eur J Soil Sci. https://doi.org/10.1111/ejss.12475

Miyata S, Kosugi K, Gomi T, Mizuyama T (2009) Effects of forest floor coverage on overland flow and soil erosion on hillslopes in Japanese cypress plantation forests. Water Resour Res 45:1–17

Molina AJ, Campo ADD (2012) The effects of experimental thinning on throughfall and stemflow: a contribution towards hydrology oriented silviculture in Aleppo pine plantations. For Ecol Manag 269:206–213

Morris BC, Bolding MC, Aust WM, McGuire KJ, Schilling EB, Sullivan J (2016) Differing levels of forestry best management practices at stream crossing structures affect sediment delivery and installation costs. Water 8(3):92. https://doi.org/10.3390/w8030092

Osterholz WR, Schwab ER, Duncan EW, Smith DR, King KW (2021) Connecting soil characteristics to edge-of-field water quality in Ohio. J Environ Qual 1–16. https://doi.org/10.1002/jeq2.20308

Owens PN, Walling DE (2002) The Phosphorus content of fluvial sediment in rural and industrialized river basins. Water research 36: 685–701. https://doi.org/10.1016/S0043-1354(01)00247-0

Pfeifer H-R, Derron M-H, Rey D, Schlegel C, Atteia O, Dalla Piazza R, Dubois J-P, Mandia Y (2000) Natural trace element input to the soil-sediment-water-plant system: examples of background and contaminated situations in Switzerland, Eastern France and Northern Italy. In: Markert B, Friese K (eds) Trace elements—their distribution and effects in the environment, vol 4. Elsevier Science B.V, Amsterdam, pp 33–86

Rehman OU, Rashid M, Kausar R, Alvi S, Hussain R (2015) Slope gradient and vegetation cover effects on the runoff and sediment yield in hillslope agriculture. Turk J Agric Food Sci Technol 3(6):478–483

Reynolds JB, Simmons RC, Burkholder AR (1989) Effects of placer mining discharge on health and food of Arctic grayling. Water Resour Bull 25:625–635

Rice RM, Wallis JR (1962) How a logging operation can affect streamflow. For Ind 89(11):38–40

Rolston DE, Harper LA, Mosier AR, Duxbury AR (1993) Agricultural ecosystem effects on trace gases and global climate change. American Agronomy Society Special Publication, Madison, p 55

Ruiz-Colmenero M, Bienes R, Marques MJ (2011) Soil and water conservation dilemmas associated with the use of green cover in steep vineyards. Soil Tillage Res 117:211–223. https://doi.org/10.1016/j.still.2011.10.004

Salome C, Coll P, Lardo E, Metay A, Villenave C, Marsden C, Blanchart E, Hinsinger P, Le Cadre E (2016) The soil quality concept as a framework to assess management practices in vulnerable agroecosystems: a case study in Mediterranean vineyards. Ecol Indic 61:456–465. https://doi.org/10.1016/j.ecolind.2015.09.047

Sensoy H, Kara O (2013) Determination of the runoff and suspended sediment from two different slope length using field plots. Artvin Coruh University J Forest Faculty 14(2):216–224

Shiklomanov IA (1998) World water resources—a new appraisal and assessment for the 21st century. A summary of the monograph world water resources. International Hydrological Programme, United Nations Educational, Scientific and Cultural Organization (UNESCO), Paris

Sthiannopkao S, Takizawa S, Wirojanagud W (2006) Effects of soil erosion on water quality and water uses in the upper Phong watershed. Water SciTechnol 53(2):45–52

Sthiannopkao S, Takizawa S, Homewong J, Wirojanagud W (2007) Soil erosion and its impacts on water treatment in the northeastern 29 provinces of Thailand. Environment International 33: 706 –711. https://doi.org/10.1016/j.envint.2006.12.007

Subramanya K (2008) Engineering hydrology, 3 edn. Tata McGraw-Hill Publishing Co. Ltd., New Delhi, p 6

Townsend KR, Pettigrove VJ, Carew MWE, Hoffmann AA (2009) The effects of sediment quality on benthic macroinvertebrates in the river Murray, Australia. Mar Freshwater Res 60(1):70–82

UNESCO (1971) Scientific framework of world water balance. International hydrological decade, United Nations Educational, Scientific and Cultural Organization (UNESCO), Paris

Wear DN, Greis JG (2002) Southern forest resource assessment. Gen. Tech. Rep. SRS-53. Southern Research Station, Forest Service, United States Department of Agriculture, Asheville, pp 635

Welde K, Gebremariam B (2017) Effect of land use land cover dynamics on hydrological response of watershed: Case study of Tekeze Dam watershed, northern Ethiopia. Int Soil Water Conserv Res 5:1–16

Winter S, Bauer T, Strauss P, Kratschmer S, Paredes D, Popescu D, Landa B, Guzmán G, Gómez JA, Guernion M, Zaller JG, Batáry P (2018) Effects of vegetation management intensity on biodiversity and ecosystem services in vineyards: a meta-analysis. J Appl Ecol 55(5):2484–2495. https://doi.org/10.1111/1365-2664.13124

Zhang L, Wang J, Bai Z, Lv C (2015) Effects of vegetation on runoff and soil erosion on reclaimed land in an opencast coal-mine dump in a loess area. CATENA 128:44–53

Zimnicki T, Boring T, Evenson G, Kalcic M, Karlen DL, Wilson RS, Zhang Y, Blesh J (2020) On quantifying water quality benefits of healthy soils. Biosci 70:343–352. https://doi.org/10.1093/biosci/biaa011

Chapter 5
Soil and Water Qualities Necessary for Irrigation

Abstract Achieving sustainability in food production needs quality soil and irrigation water. Irrigation should aim for enhancing soil resilience and ecosystem services through soil. Irrigation scheduling is very much important for meeting irrigation demand of crops, also grown in salt affected soils, for maintaining good soil and water qualities and environmental health and sustaining food production and also considering both crop and fish production from wetlands. Controlling as well as decoupling water pollution from agricultural fields and crop production as well should be considered in each crop fields for securing better human health.

Keywords Crop production · Drainage · Food production · Irrigation water · Sodium adsorption ratio · Soil · Pollution

Abbreviations

DPIRD	Dept. of Primary Industries and Regional Development
D_{dw}	Depth of drainage water
D_{iw}	Depth of irrigation water
EC	Electrical conductivity
FAO	Food and Agriculture Organisation of the United Nations
IW/CPE	Irrigation water/Cumulative Pan Evaporation
LR	Leaching requirement
PWP	Permanent wilting point
SP	Saturation % of soil
SAR	Sodium adsorption ratio
SDG	Sustainable Development Goal
UNESCO	United Nations Educational, Scientific and Cultural Organisation
USSLS	United States Salinity Laboratory Staff

5.1   Introduction

Irrigation water quality is a critical aspect of agriculture. Many factors determine water quality suitable for irrigation. The most important factors considered are soluble salts, pH and alkalinity. In addition, other important factors are hardness of water due to calcium and magnesium carbonates and factors for clogging of irrigation water supply systems. Those clogging factors are generally of three types, e.g., (1) inorganic and organic suspended solids, (2) chemical precipitation of salts of calcium, magnesium, heavy metals and fertiliser, if applied as fertigation and (3) bacterial depositions and growth of slimes and biofilms. Poor water quality causes poor growth and poor aesthetic quality of crops and may slowly lead to death of plants. Roots are injured due to presence of high concentration of soluble salts. Those salts also interfere with water and nutrient uptake by crops and cause burning of leaf margins through accumulation of those salts in plant leaf edges, thus affecting health of crops. Soil pH can be adversely affected with high alkalinity of irrigation water, and thereby, that water can interfere with nutrient uptake, resulting in nutrient deficiencies in crops and compromised plant health. Presence of disease organisms, soluble salts and traces of organic chemicals are necessary to be tested, if required, before using runoff water, reclaimed water, or recycled water for irrigation through necessary reconditioning. Analysis of water quality is needed for ensuring suitability of available water for plant growth and minimising the risk of pollutants to be discharged to surface or groundwater. So, basic understanding of soil–plant–water relationships is very important for efficiently managing cultivation of crops with irrigation and systems of irrigation and water supplies. Consequently, that work requires ascertaining suitability of land to be taken under irrigation based on detailed evaluation of topography of the land and quality of water to be used for irrigation (Ayers and Westcot 1994; Goyal et al. 2016).

5.2   Land Characterisation Necessary for Irrigation

Topography, or the 'lay of the land', is the primary concern for allowing a land to be taken under irrigation. Relief, a component of topography, is referred to the difference in elevations between higher places and depressions in the field. Topographic relief will affect various irrigation-related works on planning for the type of irrigation water conveyance (e.g., ditches or pipes) systems, assessing drainage requirements and designing water erosion control practices. Irrigation management also depends on shape, arrangement of topographic landforms and surface waterway network. For example, accumulation of water naturally after rain in the depression in the field causes those spots continually to be in wet condition with the addition of irrigation, and those wet spots may become the source of disease attack in crops like potato (Waller and Yitayew 2016).

Runoff, soil drainage, erosion, use of machinery and choice of crops are dependent on slope, i.e. the inclination or gradient of a land surface expressed in %. Shape of the slope is an important factor in choosing the type of irrigation system. Slope of the land can be classified in types of shape like plane, convex and concave; and characteristically into two types like simple and complex slopes. Slope with smooth appearance and extending in one or perhaps in two directions as found in alluvial fans and foot slopes of river valleys are regarded as simple types of slope. Complex slopes are short, concave or convex and extending in several directions most likely of tiny hill or knoll and pothole type of topography as in glacial depositional landforms or till plains. Surface (or gravity) irrigation can be followed only on simple slope of 2% or less, whereas only sprinkler or drip irrigation systems should be followed on simple and complex slopes of greater than 1%. Though centre pivot sprinkler irrigation system can operate on slopes up to 15%, but it is not generally recommended on simple slopes of greater than 9%. Land levelling and grading are of added advantage for both surface and sprinkler irrigation systems. Such operations may cause decrease in yield reductions for one to three growing seasons due to removal or disturbances caused in topsoil setting. So, special care should be taken to restore topsoil and to incorporate organic matter after such operation of land levelling (Scherer et al. 1996).

5.3 Soil and Water Compatibility Necessary for Irrigation

Planning for irrigation requires investigation on compatibility of available irrigation water with soil of the plot to be cultivated. Otherwise, incompatible water will adversely affect physical and chemical properties of the soil for cultivation of other crops in the future. In that regard, application of basic understanding of soil–plant–water relationships is required considering quality of irrigation water.

5.3.1 Interaction Between Soil and Water

Soil, being a three-phase system of solid (i.e. minerals and organic matter), liquid (i.e. water) and gases (i.e. air and moisture), retains water and soil moisture within pore spaces, comprising almost half of the volume of soil matrix. Instead of storing water, soil is also a medium of movement of water. Some of the soil physical characteristics like soil texture, bulk density and structure determine pore size and volume. In two ways, water is held in soil, viz. as thin coating on the outside of soil particles and in pore spaces. Water in soil pore spaces is held in two different forms, e.g., gravitational water and capillary water (Fig. 5.1). Gravitational water moves downward in the soil due to the force of gravity. Capillary water is held by soil particles against the force of gravity; thus, it is the most important for crop production. Soil pore spaces are filled with water through water infiltration into a soil, followed by the downward movement of water by the action of gravity and then upward water movement by capillary

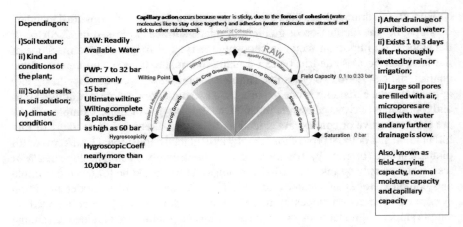

Fig. 5.1 Soil water crop relationship (based on Michael 1990; DPIRD 2019)

forces. Downward water movement continues until a balance is reached between those two opposite forces of capillarity and gravity acting on soil water. Capillary forces pull up water around soil particles through small pore spaces in any directions. Upward movement of water by capillarity from shallow water table may cause salt concentration in soil and its subsequent deposition as water is removed from soil by evaporation processes and transpiration by plants (Biswas and Mukherjee 1987; DPIRD 2019; Michael 1990).

Availability of water in the soil can be clearly described by four levels of soil moisture contents. Those four representative soil moisture levels are termed as (1) saturation, (2) field capacity, (3) wilting point and (4) hygroscopicity or oven dry condition. In the saturated soil, soil pores are filled with water, and nearly, all of the air in the soil has been displaced by water. Instantaneously, water from saturated soil will start and continue to drain for few days through the soil profile by the action of gravity, with the consequent absorption of some water by roots of plants. Water, thus, drained is called gravitational water, and after drainage of the gravitational water, a certain level of soil moisture left in the soil is termed as field capacity. So, after irrigation or rainfall, the portion of water held in the soil between saturation and field capacity is called as gravitational water. As the processes of evapotranspiration, seepage and deep drainage continue, the soil moisture content will reach to an extent when plants are unable to exert enough force to extract water from minute soil pores, and that moisture content is termed as wilting point. So, the moisture held in the soil profile between field capacity and the wilting point is available for plant use, though before reaching that point irrigator should be cautious for applying water in crop field for sustaining production. If the depletion in soil moisture content is allowed to continue, plants will be about to die and finally will not survive, and moisture content at that point is called as the permanent wilting point (PWP). With the continued reduction in soil moisture, soil moisture held in soil at about 10,000 bar atmospheric pressure is analogous to oven dry soil. That condition of soil moisture is hygroscopicity, and that moisture content is termed as hygroscopic coefficient or oven dry soil moisture which is considered as the reference for measuring other three representative soil moisture contents (DPIRD 2019; Michael 1990), as shown in Fig. 5.1.

So, maintaining the optimal quantity of irrigation water has a great bearing on the quality of soil, e.g., securing soil from salinization, waterlogging or erosion and thereby protecting both soil and water from quality deterioration.

5.3.2 *Physiological Drought Soil Condition*

Physiological drought can be defined as the inability of plants to extract water from the soil, even though it is available in the root zone. Salts in soil cause water less available to plants, then physiological drought condition also occurs due to high-soil salinity. As the salt spray (Fig. 5.2) and foam blown inland, from the sea, salt water intrusion in groundwater, capillary rise of saline groundwater and evapotranspiration are the causes soil salinity, thereby distance of inland or degree of protection from the wind are well correlated with salinity of soil and quality of inland water available for irrigation. Though the degree of attraction of water to the soil (i.e., soil moisture tension) is the most important factor of soil water characteristic for sustainable crop growth, irrigation requirement should be decided based on agroclimate-wise estimation of irrigation water/ cumulative pan evaporation (IW/CPE) ratio for a particular crop along with the consideration of suitability of salinity of soil and water and estimation of leaching requirement (LR as shown in Eq. 5.1) for flushing out salt deposit (Fig. 5.3) from soil surface and leaching below the root zone depth in the soil profile (Agarwal et al. 1982; Dhara et al. 1991; FAO and UNESCO 1967; Moreno-Casasola 2008; USSLS 1968).

$$LR = \frac{D_{dw}}{D_{iw}} = \frac{EC_{iw}}{EC_{dw}} \tag{5.1}$$

Here, leaching requirement $= LR$, i.e., water required for leaching excess salts below root zone.

LR is the ratio of depth of drainage water (D_{dw}) to depth of irrigation water (D_{iw}), which is equivalent to the ratio of electrical conductivity (EC) of iw to EC of dw (USSLS 1968).

Here,

$$D_{iw} = D_{dw} + D_{cw} \tag{5.2}$$

D_{cw} is climate-wise consumptive use of water as estimated for a crop grown, or

$$D_{iw} = D_{cw} \frac{EC_{dw}}{EC_{dw} - EC_{iw}} \tag{5.3}$$

Fig. 5.2 Salt sprays mixing in wind from sea wave break on the shore

Fig. 5.3 Salt deposition on soil surface: **a** coastal humid zone in West Bengal, India; **b** arid zone in Rajasthan, India

For sustaining crop production, saline waters can be used under suitable drainage condition. If leaching of accumulated salts not done by the use of saline water for irrigation, the established new soil salinity level would be a function of salinity of irrigation water, texture of soil and mean annual rainfall of the area of the crops grown as follows (Eq. 5.4):

$$EC_s = K + k^2 \left(\frac{EC_{iw} \times SP}{R} \right) \tag{5.4}$$

where

(1) ECs is level of salinity of soil
(2) SP is saturation % of soil and
(3) R is mean annual rainfall

(4) Two constants appear: K depending on climate, and k depending on soil and cropping pattern.

Equation (5.4), originally worked out in Israel for citrus orchards, is also used for arid and semi-arid areas of Rajasthan, North Gujarat, Haryana and Western Uttar Pradesh in India (Agarwal et al. 1982; FAO and UNESCO 1967).

5.3.3 Diagnosis of Soil Properties for Irrigation Management

Irrigation is important for securing sustainability in agricultural production. But all lands cannot be taken udder irrigation because of some soil and land-related limitations. For that purpose, numerical rating for irrigation is followed based on total 100 marks, comprising soil texture (20 marks), permeability (20 marks), total soluble salts (25 marks), pH (15 marks) and exchangeable sodium (10 marks) and noncapillary porosity (10 marks). Following that system four land classes suitable for irrigation, viz. Class I with 86 marks, no problem; Class II with 76–86 marks, moderate problems; Class III with 66–76 marks, severe problems; and Class IV with less than 66 marks, very severe problems. Another numerical method for designating five land irrigability classes of A (none to slight soil limitation), B (moderate), C (severe), D (very severe) and E (unsuited for irrigation) based on limitations w.r.t. soil depth, texture, permeability, available water holding capacity, salinity and alkalinity and drainage characteristics for continued use of agricultural lands under irrigation for sustainable crop production (Agarwal et al. 1982; Bali and Khybri 1982; Mehta and Shankaranarayana 1961; Mehta et al. 1958).

Soil structure controls movement of air, water and plant roots through a soil. That soil condition is achieved through stable aggregate with a network of soil pores for rapid exchange of air, water with crop roots, thereby controlling rate of crop growth. Grouping of particles of sand, silt and clay into larger aggregates of various sizes and shapes is the determining factor of stability of soil structure. Presence of inorganic and organic cementing agents, wetting and drying cycles, freezing and thawing, animal activity and processes of root penetration are also responsible for formation of soil structure. Structural aggregates, resistant to physical stress, are critical to the maintenance of soil tilth and productivity. Excessive cultivation or tillage of wet soils (Fig. 5.4) may cause disruption of aggregates, and loss of organic matter is accelerated, causing decrease in soil aggregate stability (Scherer et al. 1996).

Plant roots tend to be concentrated in the upper part of the soil profile due to presence of restrictive subsurface layers often interfering with root penetration. Soil depth, i.e., the thickness of the soil materials is the seat for soil structural support for securing provision of nutrients and water for plants. Presence of contrasting soil layer of sand and/ or gravel at low depth (about less than 3 feet) is responsible for decrease in available soil water for plants and thus decrease in rooting depth (Fig. 5.5). More frequent irrigations are needed for such soil conditions with less available water for plants.

Fig. 5.4 Tillage of wet soils

Fig. 5.5 Sand layers near surface causing poor rooting depth

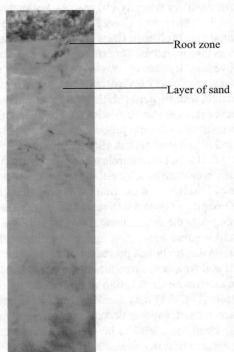

Permeability of soil is controlled by size, shape and continuity of the pore spaces and, thus, is dependent on the bulk density, structure and texture of soil. Thus, it is important for deciding irrigability of soil. Slow, very slow, rapid or very rapid permeability classifications of soils are considered as poor for irrigation. Infiltration, i.e., the downward flow of water from the soil surface, is dependent on soil surface conditions such as roughness (tillage and plant residue), plant cover, moisture content, slope and permeability. Rate of infiltration increases with increase in coarse soil texture, i.e., in the order of clay <silt <loam <sand, whereas this tendency is inverse for the stability of soil structure. For that reason, more water and nutrients are lost from soils of coarse textures. So, timing and forms of application of fertilisers are particularly important concerning soil textures (Biswas and Mukherjee 1987).

Thus, soil physical characteristics require careful observations for deciding irrigation water management in the crop fields.

5.3.4 Irrigation Water Quality

Water pollution is one of the major issues in the present situations of industrialization, and it is aggravated by the bad habits of water uses. Surface water is traditionally used for irrigation, and availability of water is affected by consumption of about 33 and 90% of the world's water sources and 70% fresh world water for this purpose. Quality of water for irrigation implies assessment of necessary physical, chemical and biological parameters of water available to determine its applicability for use in soils. Continuous monitoring of physicochemical parameters of water and soil is important for generation of useful data and information for managing and mitigating problems in water resource essential for the sustainability of crop production and productivity (Guerra Tamara et al. 2022). In the present scenario of climate change, not much information is available on the quality of the water used for irrigation, except already conventionally established findings on the effects of salinity and sodicity of irrigation water associated with high levels of electrical conductivity (EC) and a high-sodium adsorption ratio (SAR) with few usual management activities in tropical areas to cope up with unwanted soil problems.

Such problem soils with salinity and sodicity produce abiotic stress. That stress has impacts on morphological, physiological and biochemical processes, including seed germination, growth and water and mineral absorption by crops and thereby affects yields, quality and productivity of crops. So, the sustained effort is required to continuously resolve that conventional issue of problem soils. Additionally, salinity problem of soil affects the osmotic potential of soils, causing hydric tress and toxic effects on crops, exhibiting symptoms of metabolic and nutritional disorders in plants. Instead in certain conditions, soil texture may be the cause of reduced yields, lower water infiltration, crust formation on surface and clogging of pores and thereby degradation of soil and increase in runoff (Guerra Tamara et al. 2022). Irrigation water always contains some dissolved salts. So, the nature, quantity and proportion of ions present indicate the quality of irrigation water. In addition, suitability of irrigation water quality is determined by both the quantity and types of salts present (Guerra Tamara et al. 2022).

From an agricultural perspective, water quality should be analysed for (i) concentration of soluble salts, (ii) relative concentration of sodium with respect to other cations, (iii) concentration of boron or other elements that may be toxic under certain conditions and (iv) concentration of bicarbonates in relation to the concentration of calcium and magnesium (Guerra Tamara et al. 2022). Industrial discharges, use of agricultural chemicals and other chemicals and their subsequent mixing with irrigation water through runoff and infiltration of water cause significant effect on physicochemical characteristics of the water and thereby impact on sustainable soil quality and crop production. Such changes have long-term effects on soil properties which ultimately may transform the land unsuitable for agriculture (Bortolini et al. 2018; Guerra Tamara et al. 2022). Plugging of emitters and sprinklers of irrigation system is also the other ill effects of quality of irrigation water, invariably with presence of salts and sometimes with excessive salt concentrations. Presence of inorganic solids (silt and sand), organic solids (algae, bacteria and slime) and dissolved solids (calcium, iron and manganese) is also the unwanted quality of irrigation making water quality unsuitable for irrigation and creating operational problems by clogging and deposition in irrigation system.

Degree of such problems can be anticipated by testing the water quality to avoid potential complicacies in irrigation (Guerra Tamara et al. 2022). Irrigation and water resource authorities from several countries and other international organisations like the Food and Agriculture Organisation of the United Nations (FAO) have proposed the classification and monitoring methodologies as a tool to assess the suitability of water quality for irrigation (Guerra Tamara et al. 2022). Use of such tool and country-wise regulation on the irrigation water use as per international standards will be of help for countries to monitor the indicators of the Sustainable Development Goals (SDGs) associated with water quality (Guerra Tamara et al. 2022), containing the threshold values based on criteria of optimal crop yields, crop quality, suitability of the soil and maintenance of irrigation equipments (Guerra Tamara et al. 2022). In each Country, zone wise, a set of clear and precise parameters is decided to determine the maximum and minimum values of minerals, metals, pH, EC and SAR of water for irrigation. Instead that set of criteria is also helpful to supply required nutrients to both the crops and the soils for achieving good yields and productivity, as well as helping for mitigating the effects of climate change. In that context, it is considered important for establishing a baseline for the development of agricultural policies and management actions for promotion of the sustainable development and food security of small farmers (Guerra Tamara et al. 2022; Taghizadehghasab et al. 2021).

5.4 Irrigation Management in Salt Affected Soils

Salt-affected soils are classified based on their contents of soluble salts and sodium. Usually, saline and sodic soils are the results of upward movement of groundwater from a shallow water table close to the soil surface. Dissolved minerals (salts) carried by water accumulate in the soil due to evaporation of water from soil surface or transpiration by plants to the atmosphere. Those soils are not generally recommended for application of irrigation water. Though the saline and sodic soils are of natural origin, irrigation causing such problems is the cause of man-made origins of such problem soils. Certain combinations of irrigation water quality and soils may cause salts and/or sodium to accumulate in the root zone with an adverse effect on plant growth. That can be controlled by the use of soluble calcium amendments which can improve soil structure. That practice may be followed in field situations where a majority of unaffected irrigable soils contains pockets (inclusions) of sodium-affected soils and under irrigated area with surface crusting. That also requires special irrigation needs and management practices for leaching or controlling the water table elevation for managing salt concentrations. Leaching can be realised by applying more water than the soil will hold in the root zone which can be possible through large rainfall events or applying additional irrigation water or both. But that practice may cause rising up of the groundwater table and thereby capillary rise of salts again on soil surface. That dilemma may be managed by planting deep rooted plants like trees, deep rooted crops like alfalfa; installation of subsurface drainage system, where ever possible in conjunction with leaching of surface salt deposit (Agarwal et al. 1982; Biswas and Mukherjee 1987).

5.5   Diagnosis of Salt Affected Soils

For precise determination of the severity of the problem of soil salinity, salt and sodium contents of soil are required to be determined. Estimation of salt content of soil is done by measuring electrical conductivity of either a soil water extract or soil water slurry, i.e., soil paste. The sodium content of the soil is often measured on a soil water extract and is expressed as SAR, i.e., a ratio between the sodium and calcium plus magnesium contents. Monitoring that SAR value for a particular soil, based on soil sampling from surface soil layer of up to 15 cm depth periodically every three to five years, will be a good indicator to track the buildup of sodium and accumulated salt in soil and associated soil physical problems. For example, soils with SAR value of 13 will show dispersal of clay particles and the associated soil physical problems; and that value of less than SAR 6 will not have soil physical problem whereas increase in SAR values from 6 to 9 create alert for necessary preparedness for managing such unwanted soil physical problems (Agarwal et al. 1982; Bortolini et al. 2018; USSLS 1968).

5.6   Soil and Water Management for Sustainable Crop Production

Decision-making on starting irrigation and of water to be applied to crop fields is the most difficult parts of irrigation water management. Knowing the water use patterns during the different crop growth stages will aid in taking such decisions on irrigation water application, irrigation system design and management. For an improved management of water applications, it is to be noted that plants are most susceptible to damage from water deficiency during the vegetative and reproductive stages of growth by which crop water stress can be avoided and curtailment of increased pumping costs arising out of unnecessary water application which may invite more disease infections. Other criteria of improved irrigation water management should include reducing fertiliser and pesticide leaching for minimising costs of cultivation as well as protecting water resources and the environment as a whole. So, criteria of optimum application of irrigation water should be followed because too little water application at the wrong time will cause crop stress and reduced yields. In that regard, practice of irrigation scheduling should be based on accurate measurement of the rainfall received on each irrigated field by installing at least one and possibly two rain gauges (at least 5–6 cm in diameter) mounted on posts next to each irrigated field. Such practice of irrigation water management will be helpful in balancing between crop water needs, application of water at the root zone, water availability, cost of pumping, environmental health and sustaining crop production considering climatic factors and soil properties (Michael 1990, Scherer et al. 1996).

5.7   Concluding Remarks

Both soil and irrigation water qualities are very much important towards achieving sustainability in food production. Irrigation should aim for enhancing soil resilience and ecosystem services through soil as a biomembrane for sustaining water quality by denaturing contaminants, filtering and reducing nonpoint source of pollution and other pollutants. On the other hand, water qualities are also very much essential information for meeting irrigation demand of crops grown in salt affected soils for which both soil and water qualities are needed to be considered (Agarwal et al. 1982; Lal 2010; USSLS 1968). Water quality information is also important input related to fish production from wetlands (Jaishankar et al. 2014) and controlling as well as decoupling water pollution from agricultural fields and crop production as well (FAO 2013; Mateo-Sagasta et al. 2017).

References

Agarwal RR, Yadav JSP, Gupta RN (1982) Saline and Alkali soils of India. Indian Council of Agricultural Research, New Delhi

Ayers RS, Westcot DW (1994) Water quality for agriculture. FAO irrigation and drainage paper 29, Rev. 1. Food and Agriculture Organization of the United Nations (FAO), Rome

Bali YP, Khybri ML (1982) Land evaluation. In: Randhawa NS, Goswami NN, Abrol IP, Krishna Murti GSR, Ghosh AB, Prihar SS, Murthy RS, Singh S, Sastry TG, Narayanasamy G, Kumar K (eds) Review of soil research in India. ISSS/AISS/IBG 12th international congress on soil science. 8–16 Feb 1982, Part II:587–596. Industrial Soil and Soil Science, New Delhi

Biswas TD, Mukherjee S (1987) Textbook of soil science. Tata McGraw-Hill Publishing Co., Ltd., New Delhi

Bortolini L, Maucieri C, Borin M (2018) A tool for the evaluation of irrigation water quality in the arid and semi-arid regions. Agron 8(2):23. https://doi.org/10.3390/agronomy8020023

Dhara P, Panda S, Roy GB, Datta DK (1991) Effect of waterbodies on the quality of groundwater in coastal areas of South 24 Parganas in West Bengal, India. J Ind Soc Coastal Agric Res 9(1/2):395–396

DPIRD (2019) Calculating readily available water. Agriculture and Food, Department of Primary Industries and Regional Development (DPIRD), Government of Western Australia. Available: https://www.agric.wa.gov.au/citrus/calculating-readily-available-water. Accessed 21 June 2022

FAO and UNESCO (1967) International source-book on irrigation and drainage of arid lands in relation to salinity and alkalinity. Food and agriculture organization of the United Nations (FAO), Rome, and UNESCO, United Nations Educational, Scientific and Cultural Organization (UNESCO), Paris

FAO (2013) Guidelines to control water pollution from agriculture in China: decoupling water pollution from agricultural production. FAO water reports 40. Food and Agriculture Organization of the United Nations (FAO), Rome

Goyal MR, Chavan VK, Tripathi VK (2016) Principles and management of clogging in micro irrigation. In: Goyal MR (ed) Innovations and challenges in micro irrigation, vol 1. Apple Academic Press, Inc., Oakville, p 43

Guerra Tamara B, Torregroza-Espinosa AC, Pinto Osorio D, PallaresM M, Corrales Paternina A, Echeverría González A (2022) Implications of irrigation water quality in tropical farms. Global J Environ Sci Manag 8(1):75–86. https://doi.org/10.22034/GJESM.2022.01.06

Jaishankar M, Tseten T, Anbalagan N, Mathew BB, Beeregowda KN (2014) Toxicity, mechanism and health effects of some heavy metals. Interdiscip Toxicol 7(2):60–72. https://doi.org/10.2478/intox-2014-0009

Lal R (2010) Managing soil to address global issues of the twenty-first century. In: Lal R, Stewart BA (eds) Food security and soil quality. CRC Press, New York, pp 6–19

Mateo-Sagasta J, Zadeh SM, Turral H, Burke J (2017) Water pollution from agriculture: a global review—executive summary. CGIAR research program on water, land and ecosystems (WLE). Food and agriculture organization of the United Nations (FAO), Rome, and International Water Management Institute (IWMI), Colombo

Mehta KM, Mathur CM, Shankaranarayana HS (1958) A proposed method for rating lands for irrigation and its application to Chambal area. J Soil Wat Conserv 6: 125–139

Mehta KM, Shankaranarayana HS (1961) Rating of lands for irrigation - Jawai Project Area. J Indian Soc Soil Sci 9(2): 71–76

Michael AM (1990) Irrigation—theory and practice. Vikas Publishing House Pvt. Ltd. N. Delhi, pp 463, 480–482

Moreno-Casasola P (2008) Dunes. Ecosystems In: Jørgensen SE, Fath BD (eds) Encyclopedia of ecology. Imprint Elsevier Science, Elsevier B.V., Amsterdam, pp 971–976

Scherer TF, Seelig B, Franzen D (1996) Soil, water and plant characteristics important to irrigation. Revised by: Scherer TF, Franzen D, Cihacek L (2017) North Dakota State University (NDSU)

extension service, Fargo, North Dakota, USA. Revised Dec 2017. AE1675 (Revised). Available: https://www.ndsu.edu/agriculture/sites/default/files/2022-03/ae1675.pdf. Accessed 21 June 2022

Taghizadehghasab A, Safadoust A, Mosaddeghi MR (2021) Effects of salinity and sodicity of water on friability of two texturally-different soils at different matric potentials. Soil and Tillage Research 209: 104950. https://doi.org/10.1016/j.still.2021.104950

USSLS (1968) Richards LA (ed) Diagnosis and improvement of saline and alkali soils. United States Salinity laboratory staff (USSLS), agriculture handbook no. 60, USDA. Oxford & IBH Publishing Co., New Delhi

Waller P, Yitayew M (2016) Irrigation and drainage engineering. Springer International Publishing, Cham

Chapter 6
Soil Moisture Conservation Influencing Food Production

Abstract Soil moisture conservation is most useful for managing water demand by crops for a short period of drought, because soil moisture cannot be retained for long period. It is the function of soil moisture content at field capacity, bulk density of soil, effective hydrological depth or rooting depth and evapotranspiration. So, conservation of in situ rainfall as well as surface runoff in rainwater harvesting ponds and in groundwater reservoir through recharge will be helpful for supplying water for irrigation from ponds or soil moisture supply to root zone of crops as made plausible through use of mulches.

Keywords Conservation · Crop rotation · Groundwater · Harvesting · Runoff · Soil moisture · Mulch · Soil organic matter · Nature-based solutions

Abbreviations

BD	Bulk density of the soil
DSR	Direct seeded rice
EHD	Effective hydrological soil depth
E_t/E_0	Ratio of actual to potential evapotranspiration
FAO	Food and Agriculture Organization of the United Nations
MS	Soil moisture content at field capacity
Nbs	Nature based solutions
R	Rain
Ro	Mean rain per rainy day
RD	Topsoil rooting depth
Rc	Soil moisture storage capacity
SOM	Soil organic matter
SRI	System of rice intensification
UN CTCN	UN Climate Technology Centre & Network
UNFCCC	United Nations Framework Convention on Climate Change
UN	United Nations

© The Author(s), under exclusive license to Springer Nature Switzerland AG 2022   79
S. Panda, *Soil and Water Conservation for Sustainable Food Production*,
Chemistry of Foods, https://doi.org/10.1007/978-3-031-15405-8_6

6.1   Introduction

Soil moisture (or soil water), based on the concept of civil foundation engineering, is an inevitable part of the three-phase system of soil (Fig. 6.1), i.e., solid (soil minerals and organic matter), liquid (or water) and gases (soil air and vapour or soil moisture). Hence, soil moisture has significant influence on engineering, hydrological, agronomic, biological and ecological behaviours of soil. Instead soil organic matter (SOM) in soil solid is not considered in the civil foundation engineering, still that SOM has great impact on soil moisture content and soil nutrient dynamics in soil–plant–atmosphere continuum targeting sustainable food production (Arora et al. 2008; Murthy 2002; Terzaghi 1943; Venkatramaiah 2006).

   In the present situation of climate change and increase in population, intersectoral competition for land and water resources has become intense, so the scope for expansion of irrigated areas and conversion of other land to agriculture is extremely limited. Thus, this constrained situation now compels the cultivators, planners and financers to search for new innovations for nature-based solutions, i.e., suitable rainfed farming. In rainfed agriculture, conservation and management of soil moisture are very much important. To attain high agricultural productivity, soil moisture conservation is practised in various ways for reducing moisture loss from the soil. The primary goal of that practice is to reduce loss of moisture from soil through evaporation and transpiration from plant surface or reducing combined loss of moisture as evapotranspiration from cultivated plots. Soil moisture conservation is primarily considered for reducing crop irrigation requirements as per evapotranspiration and thereby maintaining the essential water supply for sustainable agricultural production. To increase the soil

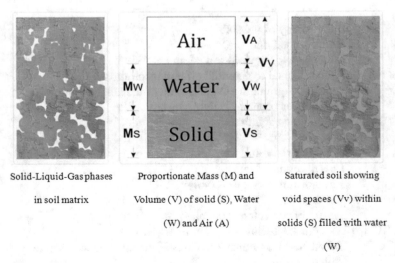

Solid-Liquid-Gas phases   Proportionate Mass (M) and   Saturated soil showing

in soil matrix   Volume (V) of solid (S), Water   void spaces (Vv) within

                                        (W) and Air (A)   solids (S) filled with water

                                                                                       (W)

Fig. 6.1   Three phases of soil: soil (S), water (W) and air (A) and their mass (M) and volume (V) relationships

moisture availability to plants from surface soil and increasing water infiltration for moisture conservation deep into the soil profile and recharge of groundwater, some site-specific recommended soil moisture conservation techniques are followed. In addition, many agronomic practices, e.g., direct seeded rice (DSR), aerobic rice, system of rice intensification (SRI), etc. are promoted to improve water productivity and water use efficiency (FAO 2021; Kumar et al. 2021; King-Okumu 2021; Midya et al. 2021a, b; Mohanta et al. 2021).

6.2 Soil Moisture Storage as Affected by Rooting Depth, Soil Bulk Density, Rainfall and Evapotranspiration

Water storage in the soil is affected by the depth and density of the roots. It is especially dependent on depth of A-horizon. Topsoil rooting depth (RD), the effective hydrological depth, is much shallower than the depth of A-horizon. The recommended values of RD are 0.05 m for grass and 0.1 m for trees and tree crops (Morgan et al. 1984) within which the storage of water affects the generation of runoff. It was assumed by Kirkby that runoff occurs when the daily rainfall exceeds the soil moisture storage capacity (Rc, mm) and that daily rainfall amounts approximate an exponential frequency distribution, and the estimation of annual runoff can be done by means of Eq. (6.1) (Kirkby 1976; Morgan RPC 2001):

$$Q = R \exp(-Rc/Ro) \qquad (6.1)$$

where:

(a) R is for rain
(b) Ro = mean rain per rainy day (mm). It is calculated as the ratio between R and the number of rainy days in the year
(c) Rc is soil moisture storage capacity in Eq. (6.1).

Rc is also estimated from Eq. (6.2):

$$Rc = 1000 \, \text{MS} \times \text{BD} \times \text{EHD}(E_t/E_o) \qquad (6.2)$$

where:

(a) MS = soil moisture content at field capacity (% w/w)
(b) BD = bulk density of the soil (Mgm^{-3})
(c) EHD = effective hydrological soil depth (m)
(d) E_t/E_0 = ratio of actual to potential evapotranspiration.

Here, EHD replaces the concept of rooting depth RD and indicates the depth of soil within which moisture storage capacity controls the generation of runoff. EHD is a function of the plant cover, and it influences the depth and density of roots and, in some instances, the effective soil depth. For example, on soils shallower than 0.1 m

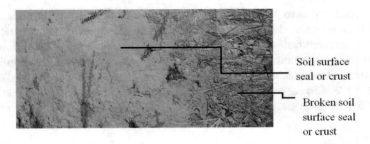

Soil surface
seal or crust

Broken soil
surface seal
or crust

Fig. 6.2   Soil surface seal or crust

or with a surface seal or crust (Fig. 6.2) (Morgan 2001, 2005; Shrestha et al. 2014; Sterk 2021).

6.3   Soil Moisture Conservation Techniques and Implementation

Soil moisture conservation techniques are important for the areas where sources of irrigation are becoming scarce, especially in the present context of uncertainly variable rainfall and scarce groundwater resources. Most of those soil moisture conservation techniques are less complex and low-cost measures and are based on locally available materials and technical capacities, i.e., good examples of nature-based solutions. For examples, provisioning some kinds of soil cover against direct soil exposure to heat and sun for minimising evaporation from soil, thereby lowering evapotranspiration. Soil and water conservation and soil quality improvement practices are generally beneficial towards soil moisture conservation (UN CTCN 2017). Some of the examples of reducing loss of soil moisture are listed as follows:

(1)   Spreading of compost or manure on soil lowers the evapotranspiration and adds nutrients to soil. Compost can be placed in the round trench under the soil surface and above the root zone depth and then covered with soil around tree (Fig. 6.3). Manure application helps in improving soil structure, soil moisture retention capacity and adding soil nutrients and thereby improving soil quality for sustaining soil fertility and productivity.

(2)   Application of mulch covering the soil surface in the low to medium rainfall areas. This is practised by placing of mulch materials like straw, wood chips, peat, plant residues like leaf litter, azolla, clover, water hyacinth, etc. (Fig. 6.4). Avoiding plastics as mulching will be helpful in creating a plastic-free environment. Application of mulch of plant residues helps in decreasing soil erosion.

(3)   Conservation tillage. It is the practice of reducing or suitably following the tillage practices. That is followed to maintain healthy soil organic matter content for

Manure
covered

Manure
applied
in a trench

Fig. 6.3 Application of organic manure around the bottom of the tree under soil surface and then covered by soil for soil moisture conservation after rainfall

Fig. 6.4 Mulching application in vegetable cultivation in a rainfed area

Mulch materials of straw and other plant residues

increasing the water absorbing and retaining capacities of soils. Through such practices, crop residue is left over in the field for reducing evapotranspiration as well as protecting soil surface from direct sun, wind and impacts of heavy rain. Mechanical tillage practices can also increase soil moisture retention. From a study, it is found that application of double-disc furrow opener with plain rolling coulter can give maximum soil moisture retention and higher bulk density with minimum soil profile disturbance and higher penetration resistance (Sawant et al. 2016). From a study on the effect of varying soil granule and degree of compaction on soil moisture loss and plant emergence in sandy soil, it is reported that on topsoil layer, better moisture retention capacity in a soil granule of

0.18 mm is found as compared to 0.76 mm soil granules (Johnson and Buchele 1961). From a study on performance of furrow openers including disc, chisel and hoe types on the effect on soil moisture and seed germination, it is observed that germination was significantly higher with chisel than hoe and disc due to more soil moisture retention achieved in chisel than hoe and disc furrow openers (Baker 1976). Operation of hoe and deep furrow openers can retain more moisture than single and double discs (Wilkins et al. 1983). More seed germination percentage is obtained by inverted-T compared to other furrow openers (Choudhary 1988; Sawant et al. 2016). It is observed that increase in speed of furrow openers causes decrease in soil moisture retention, because more soils are disturbed and exposed to wind (Tessier et al. 1991).

(4) Crop rotation by growing different types of crops every season helps in incorpo-
rating more SOM through crop residues, e.g., leaf litter, root residues, etc. and thereby can improve soil structure as well as water holding capacity of soil. Crop rotations with alternate deep and shallow rooted crops as well as cultivation of differently rooted crops help in use of water remaining at deeper depths, thus increasing water use efficiency. In this way, improvements in SOM and soil moisture status through crop rotations help in improving nutrient holding capacity, i.e., fertility of soil and helping in sustainability in food production from the same soil in future. Inclusion of green manuring crops in the crop rotation may also serve the purposes for adding SOM and nutrients and improving soil mois-
ture retention capacity, i.e., improving soil quality. Some of the important plant species for green manure are azolla fern, wild indigo (*Baptisia australis*) herb, dhaincha (*Sesbania bispinosa*), avise (*Sesbania grandiflora*), calotropis, subabul (*Leucaena leucocephala*), Glyricidia shrubs and trees like neem (*Azadirachta indica*), mahua (*Madhuca longifolia*), Karanj (*Millettia pinnata*), etc. Incorporation of green leaf manure increases water holding capacity of soil as well as decrease in soil erosion (Fig. 6.5).

Fig. 6.5 Green manuring: black gram grown under neem plantation

Fig. 6.6 Mixed cropping and interplanting: **a** mixed cropping; **b** intercropping: banana with cabbage; **c** agroforestry; **d** planting trees in crop fields; **e** sesame cultivation with young kadam (*Anthocephalus kadamba*) trees; **f** paddy field boundary plantation with trees

(5) Deep tillage. The increasing water absorption capacity of soil may be achieved though operation of deep tillage suited for some areas and soils. This can also increase porosity and permeability of soil and minimise soil erosion.

(6) Mixed cropping and interplanting of trees. Cultivating crop combinations with different planting times and different growth periods as mixed cropping and intercropping and following agroforestry by interplanting trees within crop field and on the boundary of field help in soil moisture use and retention in soil and thereby improving soil quality and sustainability in food production (Fig. 6.6).

(7) Contour ploughing. Instead of ploughing the soil up- and downward slopes, contour ploughing is done along the contour, i.e., across the slope in the crop field (Fig. 6.7). Through such ploughing, even barriers are created against the movement of water and, thus, more water is retained in the soil and runoff velocity is reduced in alternate strips (Fig. 6.8). In that way, strip cropping helps in controlling runoff and soil erosion and thereby increasing soil moisture retention and maintaining soil fertility.

(8) Rainwater harvesting. Excess rainwater, i.e., excess runoff water is collected in dug out ponds, especially on lower elevations or depressions in the crop fields. This is helpful in minimising runoff, lowering soil erosion and making provisions for multipurpose use of harvested water (Fig. 6.9). In that way, rainwater harvesting ponds, commonly called as farm ponds, can support infiltration of water into the soil, soil moisture retention, controlling flood, managing drought, increasing biodiversity, improving soil nutrients status. All those ecosystem services from farm ponds can improve microclimate and ecology in an area. But, it is important to consider planning of farm ponds based on hydrological

Fig. 6.7 Contour ploughing: ploughing across slope

Fig. 6.8 Strip cropping: mustard (erosion resisting); and cabbage and cauliflower (erosion permitting)

balance, i.e., water balance, hydrogeological conditions and social needs in the area, otherwise excessive numbers of small ponds may get dried during dry seasons. Depth of the farm ponds should also be taken into consideration, especially in coastal areas affected by salt water intrusion in groundwater, otherwise excessive depth of ponds may cause salinity in pond water due to linkage with saline groundwater below the bottom of the pond.

6.3.1 Implementation Considerations for Soil Moisture Conservation Technologies

A general assessment of four dimensions relating to adaptation and implementation of the technology is considered and that is expressed based on indicative assessment scale of 1–5 (UN CTCN 2017):

a: Rainwater harvesting pond

b: Sheep depend on farm pond water

c: Cattles depend on farm pond water

d: Natural poultry bird depends on farm pond water

e: Integrated farming based on rainwater harvesting pond

f: Fishery in rainwater harvesting farm pond

g: Fishing from farm pond

h: Animal husbandry and human habitation based on rainwater harvesting pond water

Fig. 6.9 Rainwater harvesting farm pond and its multipurpose uses

(i)   Technological maturity: 1—in early stages of research and development, to 5—fully mature and widely used. In this context, soil moisture conservation technologies can be assessed as 5, though those are very much site specific.

(ii)   Initial investment: 1—very low cost, to 5—very high-cost investment needed to implement technology. Here, those technologies may be assessed as 1, because in some interventions, collections of materials require even no costs, because those materials might be available in the farms as procured or discarded as wastes from other works.

(iii) Operational costs: 1—very low/no cost, to 5—very high costs of operation and maintenance. In this context, soil moisture conservation technologies may be assessed as 1, because those technologies are very simple for usually skilled labourers and they can operate that technical matter within their routine works.

(iv)   Implementation timeframe: 1—very quick to implement and reach desired capacity, to 5—significant investments of time required for establishing and/or reaching full capacity. In this regard, most of the soil moisture conservation technologies can be assed as 1, because most of those technologies can be implemented just after a shower of rain, while for implementation of rainwater harvesting ponds requires excavation works during dry season and collection of rainwater within dug outs during rainy season. Also, mulching can be implemented incessantly in the dry periods, because this can work as barriers of evaporation directly from soil surface.

Generally, soil moisture conservation technologies are (i) well adapted and mature technologies with (ii) very low initial investment and (iii) operational costs and require (iv) minimum implementation timeframe.

6.4 Beneficial Roles of Soil Moisture Conservation

(i)   Benefits of many soil conservation methods generally include slowing down runoff, retention of soil moisture, increasing infiltration of water, thus help in soil moisture conservation and control of soil erosion. Depending on the materials used for soil and water conservation, additionally soil nutrient conservation, weed control, soil temperature control and control of direct effects of heavy rain, wind and the sun.

(ii) Implementation of soil moisture conservation technologies facilitates active reuse of waste organic materials, thus returning plant and crop residues to the soil through decomposition, thereby reduces needs for waste management.

(iii) Socioeconomic Benefits:

    (a) Soil moisture conservation technologies are potential in reducing irrigation needs, improving soil quality and increasing crop water productivity.

    (b) Reduction in irrigation needs, in turn, is potential in reducing energy requirements for pumping of water for irrigation (UN CTCN 2017).

6.5 Prospects and Problems of Soil Moisture Conservation Techniques

6.5.1 Prospects of Soil Moisture Conservation Techniques

Improved soil moisture conservation technologies can potentially improve soil moisture retention and overall soil quality and, thus, can reduce soil degradation. Opportunities for assimilating locally available materials can considerably reduce costs and needs for extra facilities for waste recycling and management.

Many field-level soil and water conservation methods are relatively low-cost simple approaches depending on locally available materials and technical man power. Those technologies can generate income and can make synergies between alternatives adopted by local farmers, e.g., using poultry and dairy farm wastes (i.e., poultry refuse and cow dung) as composting materials (UN CTCN 2017).

6.5.2 Problems of Soil Moisture Conservation Techniques

In some situations, crop residues are not necessarily 'residues', e.g., clover or straw as those are used for fodder purposes, thus necessitating extra investments for soil moisture conservation. Such type of problem may be overcome by planning of new cycle of crops based on site-specific situation (UN CTCN 2017).

6.6 Concluding Remarks

Conservation of water (as liquid water and gaseous water, i.e., moisture) within void spaces in the soil matrix is the cheapest method. But the stored soil moisture cannot be retained for long period. It is the function of soil moisture content at field capacity, bulk density of soil, effective hydrological depth or rooting depth and evapotranspiration. So, conservation of in situ rainfall as well as surface runoff in rainwater harvesting ponds and in groundwater reservoir through recharge will be helpful for supplying water for irrigation from ponds or soil moisture supply to plant root zone as made plausible through use of mulches.

Soil physical properties, e.g., soil moisture retention, soil bulk density and soil penetration resistance decrease with increasing forward speed for all types of furrow openers. The lowest changes in soil moisture are observed in case of double-disc furrow opener with plain rolling coulter. Such attachment can open fine slit in the soil without much soil disturbances and can cut all maize stalk residue at all selected forward speeds. So, such attachment is best suited for conservation agriculture in maize–wheat cropping system.

References

Arora S, Hadda MS, Bhatt R (2008) Tillage and mulching in relation to soil moisture storage and maize yield in foothill region. J Soil Water Conserv 7(2):51–56

Baker CJ (1976) Experiments relating to techniques for direct drilling of seeds into untilled dead turf. J Agric Eng Res 21(2):133–134

Choudhary MA (1988) A new multicrop inverted T seeder for upland crop establishment. Agric Mech Asia 19(3):37–42

FAO (2021) The state of the world's land and water resources for food and agriculture—systems at breaking point. Synthesis report 2021. Food and agriculture organization of the United Nations (FAO), Rome. Available: https://doi.org/10.4060/cb7654en. Accessed 21 June 2022

Johnson WH, Buchele WF (1961) Influence of soil granule size and compaction on rate of soil drying and emergence of corn. Trans ASAE 4(2):170–174

King-Okumu C (2021) A rapid review of drought risk mitigation measures: integrated drought management. Food and agriculture organization of the United Nations (FAO), Rome. https://doi.org/10.4060/cb7085en

Kirkby MJ (1976). Hydrological slope models: the influence of climate. In: Derbyshire E Ž (Ed.) Geomorphology and Climate. Wiley, London, pp. 247–267.

Kumar PSM, Sairam M, Praharaj S, Maitra S (2021) Soil moisture conservation techniques for dry land and rainfed agriculture. Indian J Nat Sci 12(69):37386–37391

Midya A, Saren BK, Dey JK, Maitra S, Praharaj S, Gaikwad DJ, Gaber A, Alsanie WF, Hossain A (2021a) Crop establishment methods and integrated nutrient management improve: Part I. Crop performance, water productivity and profitability of rice (Oryza sativa L.) in the lower Indo-Gangetic plain, India. Agron 11:1860. https://doi.org/10.3390/agronomy11091860

Midya A, Saren BK, Dey JK, Maitra S, Praharaj S, Gaikwad DJ, Gaber A, Alsanie WF, Hossain A (2021b) Crop establishment methods and integrated nutrient management improve: Part II. Nutrient uptake and use efficiency and soil health in rice (Oryza sativa L.) field in the lower Indo-Gangetic plain, India. Agron 11:1894. https://doi.org/10.3390/agronomy11091894

Mohanta S, Banerjee M, Malik GC, Shankar T, Maitra S, Ismail IA, Dessoky ES, Attia AO, Hossain A (2021) Productivity and profitability of kharif rice are influenced by crop establishment methods and nitrogen management in the lateritic belt of the subtropical region. Agron 11:1280. https://doi.org/10.3390/agronomy11071280

Morgan RPC (2001) A simple approach to soil loss prediction: a revised morgan-morgan-finney model. CATENA 44:305–322

Morgan RPC (2005) Soil erosion and conservation. Blackwell Publishing, Oxford

Morgan RPC, Morgan DDV, Finney HJ (1984) A predictive model for the assessment of erosion risk. J Agric Eng Res 30:245–253

Murthy VNS (2002) Geotechnical engineering: principles and practices of soil mechanicsand foundation engineering. Marcel Dekker Inc., New York

Sawant S, Kumar A, Mani I, Singh JK (2016) Soil bin studies on the selection of furrow opener for conservation agriculture. J Soil Water Conserv 15(2):107–112

Shrestha DP, Suriyaprasit M, Prachansri S (2014) Assessing soil erosion in inaccessible mountainous areas in the tropics: the use of land cover and topographic parameters in a case study in Thailand. CATENA 121:40–52. https://doi.org/10.1016/j.catena.2014.04.016

Sterk G (2021) A hillslope version of the revised Morgan, Morgan and Finney water erosion model. Int Soil Water Conserv Res 9:319–332

Terzaghi K (1943) Theoretical soil mechanics. Wiley, New York

Tessier S, Saxton KE, Papendisk RI, Hyde GM (1991) Zero tillage furrow opener effects on seed environment and wheat emergence. Soil Tillage Res 21:347–360

UN CTCN (2017) Soil moisture conservation techniques. UN climate technology centre & network (UN CTCN), United Nations framework convention on climate change (UNFCCC) technology mechanism. Available: https://www.ctc-n.org/technologies/soil-moisture-conservation-techniques. Accessed 21 June 2022

Venkatramaiah C (2006) Geotechnical engineering, 2nd edn. New Age International (P) Ltd., Publishers, New Delhi

Wilkins DE, Muilenburg GA, Allamaras RR, Johnson CE (1983) Grain drill opener effects on wheat emergence. Trans ASAE 26(3):651–655

Chapter 7
Management of Soil Organic Carbon

Abstract Management of soil organic matter is essential prerequisite for assuring food security through sustainability in crop production. Soil organic matter has the ability to modify soil physical, chemical and biological properties, and thereby soil organic matter can control soil erosion by modifying soil aggregate stability, soil–water condition, soil nutrient availability, etc. Soil organic carbon is the soil health indicator which is attained through judicious application of organic matter and subsequent conversion of those materials into soil organic matter in individual crop fields for targeting sustainability in soil fertility in each crop fields and thus achieving food security.

Keywords Carbon · Food production · Humus · Microorganism · Soil organic carbon · Soil organic matter · Water

Abbreviations

AOM	Additional fraction of organic matter
C	Carbon
C_{ox}	Oxidation state
CO_2	Carbon dioxide
EHS	Extractable humic substance
FA	Fulvic acid
GHG	Greenhouse gas
H	Hydrogen
HA	Humic acid
HUM	Humus
LDN	Land degradation neutrality
LivOM	Living component of organic matter
N	Nitrogen
NH_3	Ammonia
NRCS	Natural Resources Conservation Services
NonLivOM	Nonliving component of organic matter

© The Author(s), under exclusive license to Springer Nature Switzerland AG 2022
S. Panda, *Soil and Water Conservation for Sustainable Food Production*,
Chemistry of Foods, https://doi.org/10.1007/978-3-031-15405-8_7

O$_2$	Molecular oxygen
OM	Organic matter
OR	Oxidative ratio
P	Phosphorus
S	Sulphur
SOC	Soil organic carbon
SOM	Soil organic matter
SON	Soil organic nitrogen
SDG	Sustainable development goal
UNECCE	United Nations Economic Commission for Europe
UN	United Nations
USDA	United States Department of Agriculture
WFS	Water floating substance
WSS	Water-soluble humic substance

7.1   Introduction

Soil organic matter (SOM) is defined as 'the sum of all natural and thermally altered biologically derived organic materials found in the soil or on the soil surface irrespective of its source, whether it is living or dead, or stage of decomposition, but excluding the aboveground portion of living plants'. (Baldock and Broos 2012; Tan 2014). In this regard, SOM is the product of on-site decomposition of biological matter like plant, animal and microbial residues in soil (Fig. 7.1), affecting physical, chemical and biological properties of soil and overall soil health. Composition and breakdown rate of SOM affect soil properties like soil structure, porosity, water infiltration rate, moisture holding capacity; cation exchange capacity, plant nutrient availability; biological diversity and activities of soil organisms. For sustainable food production as well as soil fertility, it is essential to maintain presence of water in soil or soil moisture and exchanges of nutrients between organic matter, water and soil. Plant nutrient cycles are broken through mining of plant nutrients through intensive crop cultivation without restoring SOM and plant nutrients, causing decline in soil fertility and break down in agroecosystem.

SOM can significantly improve capacity of soil by storing and supplying of soil available essential plant nutrients and retaining of toxic elements by chelating. SOM can also enhance resilience of soil to win over changes in soil acidity and helping minerals for faster decomposition. Soil carbon as SOM plays the central role in functioning of soils to produce a wide range of vital environmental goods and services resulting in many beneficial factors for sustainability in soil fertility and healthy ecosystem. For that reason, SOM can resolve many soil threats and those can be achieved though judicious management of soil carbon. In this way, SOM has the ability to reverse land degradation and enhancement of soil functioning. Soil carbon directly influences five essential ecosystem services delivered by soil, e.g., nutrient

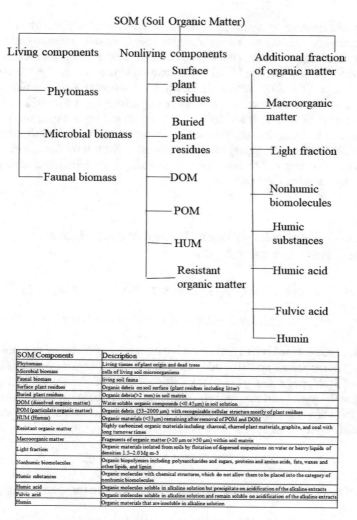

Fig. 7.1 Components of soil organic matter (SOM) (based on Baldock and Broos 2012)

cycling, water supply and quality, supporting biodiversity, sustaining soil food web and moderating climate change by reducing greenhouse gases (GHG) in the atmosphere. In that way, soil functions rendered by SOM are deciding factors in sustaining food productivity both from agricultural lands and wetlands and meet the targets of land degradation neutrality (LDN) and adaptation to changing climate (Borrelli et al. 2018; Kasimir et al. 2018; Lal 2018). Now that role of SOM as soil organic carbon (SOC) is well accepted and recognised as one of the key factors in Sustainable Development Goal (SDG) 15 of 'life on land (protect, restore and promote sustainable use of terrestrial ecosystems, sustainable management of forests, combat desertification and halt and reverse land degradation and biodiversity loss' with the Target 15.3. 'By 2030, combat desertification, restore degraded land and soil, including land affected by desertification, drought and floods and strive to achieve a land degradation-neutral world' under SDGs indicator 15.3.1 of 'proportion of land that is degraded over total land area' (UN 2018, 2022; UNECCE 2022). So, it is necessary to examine scientific outcomes on biogeochemical and physical mechanisms behind mobility of SOM as SOC for its application as information for decision-making towards mitigating and adapting to a changing climate. With the enhancement, it is required to reduce losses of SOC and to develop support mechanisms and at farm level for ensuring preservation and enhancing soil carbon stock by reversing increasing trend of soil erosion and thereby ensuring food security.

7.2   SOM, Carbon, Nitrogen, Phosphorus, Sulphur and Humus Interrelations

SOM and its components (Fig. 7.1) can be expressed through Eq. (7.1).

$$SOM = LivOM + NonLivOM + AOM \tag{7.1}$$

Here, LivOM, NonLivOM and AOM are representing (i) living components, (ii) nonliving components and (iii) additional fractions of organic matter (including humus) of SOM, respectively (Fig. 7.1). Different types of organic materials, as SOM can be generally grouped into four classes (Eq. 7.2), are (i) water floating substances (WFS), (ii) water-soluble humic substances (WSS), (iii) extractable humic substances (EHS) and (iv) residual humin (Dou et al. 2020).

$$SOM = WFS + WSS + EHS + humin \tag{7.2}$$

Out of those four components, humus (HUM) can be expressed as in Eq. (7.3):

$$Humus = WSS(part) + EHS + humin \tag{7.3}$$

where

(a)   EHS = HA + FA
(b)   HA = humic acids
(c)   FA = fulvic acids
(d)   HUM = FA + HA + humin.

Mineralisation of SOM is responsible for release of major nutrients from humic substances within soil matrix. The reversible processes of mineralisation and immobilisation result in equilibrium proportional values of carbon (C) with nitrogen (N), phosphorus (P) and sulphur (S) in recognisable ratios of C:N, C:P, C:S specific to soil type, land situation, land use land cover and climate. Those ratios, generally C:N, in other way, act as the indicators of levels of fertility of soil for better crop growth as well as sustainability in food production, because SOM acts as the seat of availability of plant nutrients in presence of congenial soil–water condition. In that way, SOM acts as the natural 'slow-release' source of plant nutrients, other than colloidal clay and small soil particles like silt, and, thus, acts in sustaining fertility of soil in the long run.

Generally, C:N in the SOM in agricultural soils varies from 8:1 to 15:1, with the median between 10:1 and 12:1; and for grasslands and forest soils mean values of C:N are about 11.8 and 12.2, respectively. That ratio in plant materials ranges from 20:1 to 30:1; and in farm compost (Fig. 7.1), C:N ratio is as high as 100:1 (Baldock and Broos 2012; Buckman and Brady 1974). Estimation of SOM is made by considering gravimetric contents of organic carbon, and/or total nitrogen for which traditional factor of 1.72, expressed on a mass basis and assuming a carbon content of approximately 58% by weight in SOM, is used to convert gravimetric SOC (g Ckg^{-1} of soil) content into SOM (gkg^{-1} of soil), as shown in Eq. (7.4):

$$SOM = 1.72 \times SOC \tag{7.4}$$

and considering C: N ratio and soil organic nitrogen (SON, g Nkg^{-1} of soil), Eq. (7.4) can be written as in Eq. (7.5):

$$SOM = 1.72 \times C : N \text{ ratio} \times SON \tag{7.5}$$

Molar C:N:P:S ratio of SOM is estimated to be about 107:7.7:1:1; and average C:N:P ratios for soil and microbial biomass are about 186:13:1 and 60:7:1, respectively; and tendencies of grassland soils are to be more nutrient rich and forest soils to be C-rich. Though molar C:O: hydrogen (H) ratio of SOM is not generally used, another useful property of SOM, i.e., oxidation state (C_{ox}) can be calculated from its molar elemental composition. The value of C_{ox} for a given organic molecule or SOM, with x, y, z and w molar %ages for C, H, O and N, respectively, in the molecule or weighted average molecule of $C_xH_yO_zN_w$, can be calculated following Eq. (7.6):

$$C_{ox} = \frac{2z - y + 3w}{x} \tag{7.6}$$

With regard to consumption of molecular oxygen (O_2) and consequent carbon dioxide (CO_2) release during decomposition (i.e., mineralisation) process of organic matter, another most useful consideration is oxidative ratio (OR) which is defined as the molar ratio of O_2 consumption to CO_2 emission during decomposition. Based on the assumption that net photosynthesis occurs via Eq. (7.7) with all N being obtained as ammonia (NH_3), OR can be calculated from C_{ox} following Eq. (7.8):

$$xCO_2 + \frac{1}{2}(y-3w)H_2O + wNH_3 \rightarrow C_xH_yO_zN_w + [x + 1/4(y - 3w) - z/2]O_2 \tag{7.7}$$

$$OR = 1 - \frac{1}{4}C_{ox} \qquad (7.8)$$

C_{ox} values may vary from -4 to $+4$ but in organic molecules usually in between -2.2 and $+3$, e.g., C_{ox} values of -2 to -1, 0 and 0 to $+3$ for lipids, carbohydrates and organic acids, respectively. In that way, C_{ox} values of SOM and its components can be used to know their molecular composition, pathway of biochemical synthesis, level of decomposition and diagenetic background. C_{ox} values of < -1 of an organic material, for example, most likely indicate prevalence of lipid/aliphatic structures, whereas those values of > 1 are likely characteristic of high content of carboxylic acids. C_{ox} values between -1 and $+1$ are not so useful because a large range of combinations of different compounds can be tested to give such values. Such determinative values of C_{ox} are, in that way, can be applied for meaningful quantitation and use of SOM towards sustainable food production and other suitable applications in assuring ecosystem services of soil and water and overall environment (Baldock and Broos 2012).

7.3   Role of Soil Organic Matter on Soil Aggregate Stability

SOM is the medium through which life is set up in the barren mass of fractured or weathered rock particles. SOM renders ecosystem services of (i) soil conditioning, (ii) nutrient pool, (iii) supporting microbial activity as substrate, (iv) conservation of environment and (v) major determinant for sustaining and increasing agricultural productivity as well as food security. Presence of water is essential for effective performance of all those ecosystem services of soil. Soil is composed mainly of four parts of which SOM and mineral soil particles are the two portions of solids, and they constitute about half of the total volume of soil in the shares of about 5 and 45%, respectively, and another half of the soil volume is filled with water and air in equal shares as displayed in Fig. 7.2 (Buckman and Brady 1974; Kalev and Toor 2018; USDA NRCS 2022). Though in typical agricultural soils, SOM content ranges from 1 to 5%, and SOM may range from 1 to 100% from desert soils to organic soils. Low-SOM levels in soils contribute significantly in the increased bulk density values while increased bulk density will attribute to degradation of soil structure within soil profile as evidenced from tropical soils. Significantly, low values of SOM in the soil are caused by high-SOM mineralization due to high-atmospheric temperature and over use of the soil due to tillage. High-bulk density also affects movement of water in soil by developing of soil compaction and sometimes developing hardpan or impervious layers in soils (Akamigbo and Igwe 1990; Igwe 2003; Igwe et al. 1995, 2005; Mbagwu et al. 1983; Schnitzer 2005).

Stable aggregate potentials of tropical soils are high and are characterised with low-dispersion ratio and highly water stable aggregate % age due to presence of slowly expandable minerals like kaolinite, oxides and quartz when wet. In addition, presence of SOM, like sesquioxides, acts as cementing agent, increases binding of

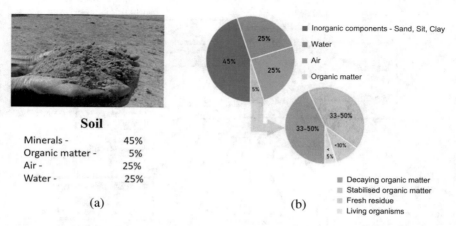

Soil

Minerals -	45%
Organic matter -	5%
Air -	25%
Water -	25%

(a)   (b)

Fig. 7.2 Composition of soil (based on Buckman and Brady 1974, Kalev and Toor 2018, USDA NRCS 2022): **a** soil only; **b** soil and SOM

soil particles in soil aggregates and reduces soil dispersion ratio, thus protecting soil from erosion (Glinski et al. 2011).

Residues of plants, animals, soil biota and microorganisms are the primary sources of all organic C-containing compounds in the soil, i.e., SOM (Fig. 7.3). Accumulation of SOM is influenced by quantity of organic residues, climatic condition, land situation, soil texture, degree of presence of water and microorganisms in the soil environment including soil redox conditions and other soil chemical and physical properties (Medina-Sauza et al. 2019; Pett-Ridge et al. 2021; Powlson et al. 2022; Schmidt et al. 2002; Stevenson 1964; Sun and Ge 2021; Wertz et al. 2012).

Humic substances are most abundant natural organic macromolecules and the major components of SOM in well humified soils. Extracts of soil humic substances are mostly partitioned into three fractions: (1) HA fraction coagulates when the extract is acidified to pH = 2; (2) FA fraction remains in solution after acidification, and it is soluble in both alkali and acid and (3) humin fraction remains with inorganic soil constituents, and it is insoluble in both alkali and acid. The structures of those humic substances are far more complex than any biopolymers synthesised under biological or genetic control. Those three fractions are not distinct chemical substances and consist of hundreds of compounds associated at molecular levels by mechanisms not clearly understood. The structures and composition of those substances are complex molecules of major naturally occurring polymers, such as proteins, polysaccharides and nucleic acids (Chen 1998; Christensen and Johnston 1997; Gerke 2018; Schnitzer 1995, 2000, 2005; Trivedi et al. 2018).

Formation of humic substances from variety of precursors in presence of heterogeneous substrates, soil biota and soil environments involves various degradative and synthetic processes. For this reason, any two humic molecules in any batch are exactly same for which it is difficult to fractionate such gross mixture into homogeneous components for detailed studies of their structures. It is conspicuous to identify

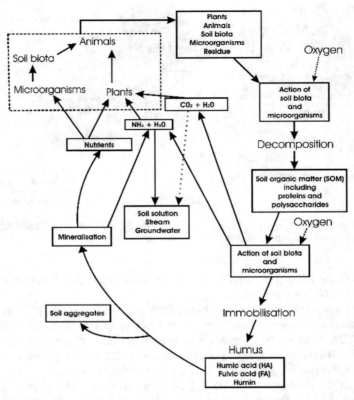

Fig. 7.3   Soil organic matter (SOM) in soil ecological cycle involving various soil functions (based on Stevenson 1964)

individuality of each humic component in each environment distinguishable from other components from the same environment and same component in different environments. Humification processes include selective preservation of refractory biological residues or components and some of their direct transformations in to humic macromolecules including synthesis of microorganisms proliferating on those organic residues. There may occur chemical synthesis of humic macromolecules, viz. widely favoured Maillard or Browning reaction involving sugar amine condensation followed by formation of Schiff base from carbonyl moiety of sugar molecule in a series of reversible reactions with an amine, typically an amino acid or lysine residue or protein (Fig. 7.4). This is the result of nucleophilic attack by the amino group of the amino acid on the electrophilic carbonyl of sugar. The formation of colour in the mixtures of aqueous reducing sugar/amino acid directly varies with the % age of reducing sugar in the aldehyde form. In addition, lignin residues can contribute to humic substances, and lignin-type structures are characteristics of newly formed humic components (Haider et al. 1965; Hopkins and Shiel 1991; Kogel-Knabner et al.

Maillard or Browning reaction

I. Sugar amine condensation:

aldose sugar + amino compound
$$\leftrightarrows N - substituted\ glycosylamine\ +\ H_2O$$

$\Leftrightarrow$ **Series of reversible reactions:**

II. Formation of Schiff base:

from carbonyl moiety of sugar molecule with an amine,
typically, an amino acid or lysine residue or protein

Fig. 7.4 Maillard or browning reaction

1992; Schnitzer 2005; Haider and Guggenberger 2005; Hayes et al. 2017; Kopecký et al. 2022).

With the progress of humification process, there decreases the addition of phenolic groups, as those are in lignin, to the humic structures. This is described as lignin mineralization, and it is accelerated in soils subjected to bare fallow and single crop culture. Such soils show total C and N less than for those under crop rotation and/or with addition of organic manures (Baldock et al. 1989; Haider et al. 1965; Hayes 2003; Kinesch et al. 1995; Pfeffer and Gerasimowicz 1989; Wilson 1987). In addition to lignin, other major components of SOM are soil polysaccharides which have important roles in soil environment. It is found that SOM fractions with different soil size fractions of same soil are different. Contents of phenol, probably of lignin origin, increase in smaller soil particles while oxygen-alkyl molecules, presumably of carbohydrate origin, are associated with largest fractions of soil particles. Due to susceptibility of SOM to decomposition and mineralisation processes, biologically labile C can be quantified by estimating C and N associated with soil microbial biomass and CO_2-C mineralised over a defined time and similarly measuring S and P also. It is also to be noted that soil architectural and structural conditions can significantly control biological decomposition of SOM affecting soil functions of water and oxygen availability, soil aggregation dynamics, accessibility of SOC to decomposer microorganisms and then microorganisms to their faunal predators. This is due to the ability of clays to confine SOM and confinement of SOC within soil aggregates and in small soil pores. As soil pores of size < 3 μm are not accessible to microbial decomposers, SOM adsorbed on clay particles in such pore conditions are decomposed as a result of diffusion of extracellular enzymes from microorganisms and then diffusion of enzymatic products back to microorganisms. Thus, soil particles can result in slow decomposition of SOM and thereby slowing down the nutrient availability in soil solution and subsequent leaching of nutrients from soil matrix. In addition, C is also retained through synthesis in microbial cells, i.e., called as 'microbial efficiency' which varies for different microbial species, soil properties and qualities of decomposing residues. Mineralisation of SOC is variously called as 'soil respiration', 'basal respiration' or 'microbial respiration' and is an indicator of biological activity in soil matrix. It is measured over periods from few days to few weeks; and assessing the release of CO_2-C on long timeframe (>3 months) is considered as a measure of SOC fraction readily available to decomposer soil organisms (Baldock and Broos 2012; Baldock and Nelson 2000; Christensen and Johnston 1997).

7.4   Concluding Remarks

From the point of view of SOC management, management of SOM is essentially prerequisite for assuring food security through sustainability in crop production. SOC is the soil health indicator which is attained through judicious application of organic matter (OM) and subsequent conversion of those OM into SOM. SOM has the ability to modify soil physical, chemical and biological properties, and thereby SOM can control soil erosion by modifying soil aggregate stability, soil–water condition, soil nutrient availability, etc. All those characteristics of SOM can assure targeting sustainability in soil fertility in each crop fields and, thus, achieving food security.

References

Akamigbo FOR, Igwe CA (1990) Physical and chemical characteristics of four gullied soils locations in Anambra State, Nigeria. Niger Agric J 25:29–48

Baldock JA, Broos K (2012) Soil organic matter. In: Huang PM, Li Y, Sumner ME (eds) Handbook of soil sciences: properties and processes, 2nd edn. CRC Press, Boca Raton, pp 1–52

Baldock JA, Nelson PN (2000) Soil organic matter. In: Sumner ME (editor-in-chief) Handbook of soil science. CRC Press, Boca Raton, pp 825–884

Baldock JA, Oades JM, Vassallo AM, Wilson MA (1989) Incorporation of uniformly labelled 13C-glucose into the organic fraction of a soil. Carbon balance and CP/MAS 13C NMR measurements. Aust J Soil Res 27:725–746

Borrelli P, Panagos P, Lugato E, Alewell C, Ballabio C, Montanarella L, Robinson DA (2018) Lateral carbon transfer from erosion in noncroplands matters. Global Change Biol 24(8):3283–3284. https://doi.org/10.1111/gcb.14125

Buckman HO, Brady NC (1974) The soil in perspective. In: Buckman HO, Brady NC (eds) The nature and properties of soils. Eurasia Publishing House, New Delhi, p 13

Chen Y (1998) Electron microscopy of soil structure and soil components. In: Huang PM, Senesi N, Buffle J (eds) Structure and surface reactions of soil particles. Wiley, Chichester, pp 155–182

Christensen BT, Johnston AE (1997) Soil organic matter and soil quality—lessons learned from long-term experiments at Askov and Rothamsted. In: Gregorich EG, Carter MR (eds) Developments in soil science: soil quality for crop production and ecosystem health. Elsevier B.V., Amsterdam, pp 399–430. https://doi.org/10.1016/S0166-2481(97)80045-1

Dou S, Shan J, Song X, Cao R, Wu M, Li C, Guan S (2020) Are humic substances soil microbial residues or unique synthesized compounds? A perspective on their distinctiveness. Pedosphere 30(2):159–167. https://doi.org/10.1016/S1002-0160(20)60001-7

Gerke J (2018) Concepts and misconceptions of humic substances as the stable part of soil organic matter: a review. Agron 8(5):76. https://doi.org/10.3390/agronomy8050076

Glinski J, Horabik J, Lipiec J (2011) Encyclopedia of agrophysics. Springer, Dordrecht

Haider K, Frederick LR, Flaig W (1965) Reactions between amino acid compounds and phenols during oxidation. Plant Soil 22:49–64

Haider KM, Guggenberger G (2005) Organic matter: genesis and formation. In: Hillel D (ed) Encyclopedia of soils in the environment, vol 3. Academic Press, Boston, pp 93–101

Hayes MHB (2003) Advances in our understanding of the composition and structures of soil organic matter. In: Wilson WS, Gray TRG, Greenslade DJ, Harrison RM, Hayes MHB (eds) Advances in soil organic matter research: the impact on agriculture and the environment. Woodhead Publishing Ltd., Cambridge, pp 1–2

Hayes MHB, Mylotte R, Swift RS (2017) Humin: its composition and importance in soil organic matter. In: Sparks DL (ed) Advances in agronomy. Academic Press, Burlington, pp 47–138

Hopkins DW, Shiel RS (1991) Spectroscopic characterization of organic matter from soil with mull and mor humus forms. In: Wilson WS (ed) Advances in soil organic matter research: the impact on agriculture and the environment. Royal Society of Chemistry, Cambridge, pp 71–90

Igwe CA (2003) Erodibility of soils of the upper rainforest zone, south eastern Nigeria. Land Degrad Develop 14:323–334

Igwe CA, Akamigbo FOR, Mbagwu JSC (1995) Physical properties of soils of south eastern Nigeria and the role of some aggregating agents in their stability. Soil Sci 160:431–441

Igwe CA, Zarei M, Stahr K (2005) Mineral and elemental distribution in soils formed on the river Niger floodplain, Eastern Nigeria. Aust J Soil Res 43:147–158

Kalev SD, Toor GS (2018) The composition of soils and sediments. In: Török B, Dransfield T (eds) Green chemistry—an inclusive approach. Elsevier Inc., Amsterdam, pp 339–357

Kasimir A, He H, Coria J, Norden A (2018) Land use of drained peatlands: greenhouse gas fluxes, plant production, and economics. Global Change Biol 24(8):3302–3316. https://doi.org/10.1111/gcb.13931

Kinesch P, Powlson DS, Randall EW (1995) 13C NMR studies of soil organic matter in whole soils: I. Quantitation possibilities. Europ J Soil Sci 46:125–138

Kogel-Knabner I, de Leeuw JW, Hatcher PG (1992) Nature and distribution of alkyl carbon in forest soil profiles: implications for the origin and humification of aliphatic biomacromolecules. Sci Total Environ 117(118):175–185

Kopecký M, Kolář L, Perná K, Váchalová R, Mráz P, Konvalina P, Murindangabo YT, Ghorbani M, Menšík L, Dumbrovský M (2022) Fractionation of soil organic matter into labile and stable fractions. Agron 12(1):73. https://doi.org/10.3390/agronomy12010073

Lal R (2018) Digging deeper: a holistic perspective of factors affecting soil organic carbon sequestration in agroecosystems. Global Change Biol. https://doi.org/10.1111/gcb.14054

Masiello CA, Gallagher ME, Randerson JT, Deco RM, Chadwick OA (2008) Evaluating two experimental approaches for measuring ecosystem carbon oxidation state and oxidative ratio. J Geophys Res 113: G03010. https://doi.org/10.1029/2007JG000534

Mbagwu JSC, Lal R, Scott TW (1983) Physical properties of three soils in southern Nigeria. Soil Sci 136:48–55

Medina-Sauza RM, Álvarez-Jiménez M, Delhal A, Reverchon F, Blouin M, Guerrero-Analco JA, Cerdán CR, Guevara R, Villain L, Barois I (2019) Earthworms building up soil microbiota, a review. Front Environ Sci 7:81. https://doi.org/10.3389/fenvs.2019.00081

Pett-Ridge J, Shi S, Estera-Molina K, Nuccio E, Yuan M, Rijkers R, Swenson T, Zhalnina K, Northen T, Zhou J, Firestone MK (2021) Rhizosphere carbon turnover from cradle to grave: the role of microbe–plant interactions. In: Gupta VVSR, Sharma AK (eds) Rhizosphere biology: interactions between microbes and plants, rhizosphere biology. Springer Nature Singapore Pte Ltd., pp 51–73. https://doi.org/10.1007/978-981-15-6125-2_2

Pfeffer PE, Gerasimowicz WV (eds) (1989) Nuclear magnetic resonance in agriculture. CRC Press, Boca Raton

Powlson DS, Poulton PR, Glendining MJ, Macdonald AJ, Goulding KWT (2022) Is it possible to attain the same soil organic matter content in arable agricultural soils as under natural vegetation? Outlook Agric 51(1):1–14. https://doi.org/10.1177/00307270221082113

Schmidt I, Sliekers O, Schmid M, Cirpus I, Strous M, Bock E, Kuenen JG, Jetten MSM (2002) Aerobic and anaerobic ammonia oxidizing bacteria competitors or natural partners? FEMS Microbiol Ecol 39:175–181

Schnitzer M (1995) Organic–inorganic interactions in soils and their effect on soil quality. In: Huang PM, Berthelin J, Bollag J-M, McGill WB, Page AL (eds) Environmental impacts of soil component interactions. Lewis Publishers, Boca Raton, pp 3–19

Schnitzer M (2000) A lifetime perspective on the chemistry of soil organic matter. Adv Agron 68:1–58

Schnitzer M (2005) Organic matter: principles and processes. In: Hillel D (ed) Encyclopedia of soils in the environment, vol 3. Academic Press, Boston, pp 85–92

Stevenson IL (1964) Biochemistry of soil. In: Bear FE (ed) Chemistry of the soil, 2nd edn. Oxford & IBH Publishing Co., New Delhi, p 258

Sun Y-Q, Ge Y (2021) Temporal changes in the function of bacterial assemblages associated with decomposing earthworms. Front Microbiol 12:682224. https://doi.org/10.3389/fmicb.2021.682224

Tan KH (2014) Humic matter in soil and the environment—principles and controversies. CRC Press, Taylor & Francis Group, 6000 Broken Sound Parkway NW, Suite 300, Boca Raton, FL 33487–2742, USA

Trivedi P, Wallenstein MD, Delgado-Baquerizo M, Singh BK (2018) Microbial modulators and mechanisms of soil carbon storage. In: Singh BK (ed) Soil carbon storage modulators, mechanisms and modeling, Chap. 3. Academic Press, Boston, pp 73–115. https://doi.org/10.1016/B978-0-12-812766-7.00003-2

UNECCE (2022) UNSDG Indicator 15.3.1: proportion of land that is degraded over total land area, (%). United Nations economic commission for Europe (UNECCE). Available: https://w3.unece.org/SDG/en/Indicator?id=66. Accessed 21 June 2022

UN (2018) High-level political forum goals in focus—goal 15: protect, restore and promote sustainable use of terrestrial ecosystems, sustainably manage forests, combat desertification, and halt and reverse land degradation and halt biodiversity loss. Statistical Division, Department of Economic and Social Affairs, United Nations (UN), New York. Available: https://unstats.un.org/sdgs/report/2018/Goal-15. Accessed 21 June 2022

UN (2022) Sustainable development goal 15. Department of Economic and Social Affairs, United Nations (UN), New York. Available: https://sdgs.un.org/goals/goal15. Accessed 21 June 2022

USDA NRCS (2022) Soil organic matter: soil quality kit—guides for educators. Natural services conservation services (NRCS), United States Department of Agriculture (USDA), Washington, DC. Available: https://www.nrcs.usda.gov/Internet/FSE_DOCUMENTS/nrcs142p2_053264.pdf. Accessed 21 June 2022

Wertz S, Adam KK, Leigh AKK, Grayston SJ (2012) Effects of long-term fertilization of forest soils on potential nitrification and on the abundance and community structure of ammonia oxidizers and nitrite oxidizers. FEMS Microbiol Ecol 79:142–154. https://doi.org/10.1111/j.1574-6941.2011.01204.x

Wilson MA (1987) N.M.R. techniques and applications in geochemistry and soil chemistry. Pergamon Press, Oxford

Chapter 8
Concluding Remarks: Soil and Water for Food Security

Abstract The United Nations Sustainable Development Goal (SDG 15) for sustainable food production can be assured by plot-wise management of soil erosion, soil organic matter, soil moisture, irrigation water, soil salinity, mulching application, growing cover crops and agroforestry in each farm. Such farming practices would result in regional as well as country-level cumulative impacts on good outcomes of applications of plot-level soil water conservation measures in each crop fields. This needs back-up from continuous studies, research, education in soil and water conservation as a full-fledged stream of science and extension services from soil water conservation scientific community as per societal needs and acceptability.

Keywords Crop production · Food production · Irrigation · Soil organic carbon · Soil organic matter · Soil water content · Sustainable development goal

Abbreviations

FAO	Food and Agriculture Organization of the United Nations
NPS	Nonpoint source
NREL	Natural Resources Ecology Laboratory
OM	Organic matter
SOC	Soil organic carbon
SOM	Soil organic matter
SWC	Soil water content
UN SDG	United Nations Sustainable Development Goal
WWC	World Water Council

8.1 Concluding Remarks

In the last century, the idea of soil resource protection has become more widespread. That has raised question on acceptance of the concepts of 'soil health', 'soil fertility' and 'soil conservation' in scientific literature and into public awareness. From a historical analysis, it is found that 'soil science community is slow to adopt those terms, but the concepts gain momentum over time' (Mizuta et al. 2021; SWC 2022). In some places, agricultural science community is still now hesitant to continue research, studies and education in soil conservation as a full-fledged stream of science.

The knowledge of soil science has grown up to the unthinkable dimensions from studies on massive soil mass to soil particles and colloidal clay micelle, soil microbial community, soil organic matter, soil nutrients, soil moisture and their uptake by plants, and from soil physical, chemical, microbial and biochemical properties and their interactions as associated with climate and hydrological conditions for successful crop cultivation and thereby application of methodologies for determining all those vast characteristics of soil are also of innumerable expanse from gravimetric methods to applications of nuclear and nanotechnologies.

Application of soil and water conservation methods also ranges from small-scale applications to big projects. Experiences on depending on big projects (like big dams) have fetched very much unsatisfactory outcomes and taking back the development to primitive conditions (like uncontrolled recurrent floods and drought) impairing the target of sustainable crop production. In that back drop of past experiences, awareness is required to be generated among policy-makers and authorities for governance of soil and water resources based on scientific logic and usefulness for implementing small-scale measures of soil and water conservation for soil health towards sustainable food production to support a global population of 9 to 10 billion (Connel 2013; FAO 2011, 2021; FAO and WWC 2015; Ghosh 2014; Luino et al. 2014; NREL 2021).

Humans depend on healthy soil to support and regulate the provision of ecosystem services of soil, including provision of food. Soil erosion, one of the global environmental problems, can lead to land degradation as well as reducing ecological service levels and on-site impact on biodiversity loss and thereby causing siltation of reservoir/lake and downstream of river. Thus, those processes cause off-site impact on endangering regional sustainable development. So, soil erosion assessment should be based on field investigation, high-resolution remote sensing and/or detailed cartographic data and the soil erosion information finally obtained at the individual field scale. That information will be helpful in planning for suitable soil and water conservation measures at farmers' field level (Amundson et al. 2015; Borrelli et al. 2017, 2021, FAO 2019a, b; Montanarella 2015; Morgan 2005; Oldeman et al. 1991, 1994; Robinson et al. 2017; Teng et al. 2018; Yang et al. 2020).

Soil quality evaluations at the farm, watershed, county, state, regional or national scales are more general and less precise than those made at the point or plot scale (Coyne et al. 2022; Creamer et al. 2022; Karlen et al. 1998; Lehman et al. 2015). So, greater efforts are required to explore individual and interactive effects of drivers of global change under controlled environment and long-term research experiments to undertake climate-adaptive strategies (Allen et al. 2011; Costantini and Mocali 2022; Hassan et al. 2022; Huang et al. 2022) involving suitable soil and water conservation measures on farm land, especially on individual plot for assuring sustainable crop cultivation and food security.

Soil erosion is the prime factor of sustained water quality. Vegetative cover may be useful in conserving soil from erosion losses, though it depends on local societal needs. Forest operations like logging, transport, construction of roads, etc. should consider careful works with minimum disturbances in the environment. In agricultural fields, vegetative cover, so far accepted by the farmer, may be practised which will be helpful in conserving loss of soil from individual crop fields. As the eroded soil take away nutrients and agrochemicals with it, conserving plot-wise soil loss has the cumulative effect on sustaining water quality and healthy environment.

Qualities of both soil and irrigation water are very much important towards achieving sustainability in food production. Irrigation should aim at enhancing soil resilience and ecosystem services through soil as a biomembrane for sustaining water quality by denaturing contaminants, filtering and reducing nonpoint source (NPS) pollution and other pollutants. On the other hand, water qualities are also very much essential information for meeting irrigation demand of crops grown in salt-affected soils for which both soil and water qualities are needed to be considered (Agarwal et al. 1982; Lal 2010; USSLS 1968). Water quality information is also important input related to fish production from wetlands (Jaishankar et al. 2014) and controlling as well as decoupling water pollution from agricultural fields and crop production as well (FAO 2013; Mateo-Sagasta et al. 2017).

Conservation of soil moisture within void spaces in the soil matrix is the cheapest method. But the stored soil moisture cannot be retained for long period. It is the function of soil moisture content at field capacity, bulk density of soil, effective hydrological depth or rooting depth and evapotranspiration. So, conservation of *in situ* rainfall as well as surface runoff in rainwater harvesting ponds and in groundwater reservoir through recharge will be helpful for supplying water for irrigation from ponds or soil moisture supply to plant root zone as made plausible through use of mulches. Regarding application of tillage machineries, it is found that soil physical properties, e.g., soil moisture retention, soil bulk density and soil penetration resistance decrease with increase in forward speed for all types of furrow openers. The lowest changes in soil moisture are observed in case of double-disc furrow opener with plain rolling coulter. Such machinery attachments, without much soil disturbances, are best suited for conservation agriculture.

Management of soil organic matter (SOM) is essentially prerequisite for managing soil organic carbon (SOC) in cultivated soil for assuring food security through sustainability in crop production. SOC is the soil health indicator which is attained through judicious application of organic matter (OM) and subsequent conversion of those OM into SOM. SOM has the ability to modify soil physical, chemical and biological properties, and thereby SOM can control soil erosion by modifying soil aggregate stability, soil–water condition, soil nutrient availability, etc. All those characteristics of SOM can assure targeting sustainability in soil fertility in each crop fields and, thus, achieving food security.

It is very much important to note that public awareness is in favour to share their patience to know soil water conservation (SWC) science and technologies, but most of those technologies are not so much adapted to societal needs as per in harmony with the locally available skills and resources as applicable to individual crop fields.

To cater to such needs of the farming community, SWC skilled scientific community needs to be employed to render their services in research, studies, education in soil and water conservation as a full-fledged stream of science and work in individual crop fields. Such approach on application of soil and water conservation science and technologies in individual farms and cropping plots will assure the nation-wise UN SDG 15 for sustainable food production and food security.

References

Agarwal RR, Yadav JSP, Gupta RN (1982) Saline and Alkali soils of India. Indian Council of Agricultural Research, New Delhi

Allen DE, Singh BP, Dalal RC (2011) Soil health indicators under climate change: a review of current knowledge. In: Singh BP, Cowie AL, Chan KY (eds) Soil health and climate change: an overview. Change. Springer, Heidelberg, pp 25–45

Amundson R, Berhe AA, Hopmans JW, Olson C, Sztein AE, Sparks DL (2015) Soil and human security in the 21st century. Sci 348(6235):1261071–1261071e1261071–1261076

Borrelli P, Robinson DA, Fleischer LR, Lugato E, Ballabio C, Alewell C, Meusburger K, Modugno S, Schütt B, Ferro V, Bagarello V, Van Oost K, Montanarella L, Panagos P (2017) An assessment of the global impact of 21st century land use change on soil erosion. Nat Comm 8(1):1–13

Borrelli P, Alewell C, Alvarez P, Anache JAA, Baartman J, Ballabio C, Bezak N, Biddoccu M, Cerdà A, Chalise D, Chen S, Chen W, De Girolamo AM, Gessesse GD, Deumlich D, Diodato N, Efthimiou N, Erpul G, Fiener P, Freppaz M, Gentile F, Gericke A, Haregeweyn N, Hu B, Jeanneau A, Kaffas K, Kiani-Harchegani M, Lizaga Villuendas I, Li C, Lombardo L, López-Vicente M, Lucas-Borja ME, Märker M, Matthews F, Miao C, Mikoš M, Modugno S, Möller M, Naipal V, Nearing M, Owusu S, Panday S, Patault E, Patriche CV, Poggio L, Portes R, Quijano L, Reza Rahdari M, Renima M, Ricci GF, Rodrigo-Comino J, Saia S, Nazari Samani A, Schillaci C, Syrris V, Kim HS, Noses Spinola D, Tarso Oliveira P, Teng H, Thapa R, Vantas K, Vieira D, Yang JE, Yin S, Zema DA, Zhao G, Panagos P(2021) Soil erosion modelling: a global review and statistical analysis. Sci Total Environ 780:146494

Connel D (2013) The tennessee valley authority: catchment planning for social development. Part of an 11-part series titled 'International water politics', Global Water Forum (https://globalwat erforum.org/). Available: https://globalwaterforum.org/2013/03/20/international-water-politics-the-tennessee-valley-authority-catchment-planning-for-social-development/. Accessed 20 June 2022

Costantini EAC, Mocali S (2022) Soil health, soil genetic horizons and biodiversity. J Plant Nutr Soil Sci 185:24–34

Coyne MS, Pena-Yewtukhiw EM, Grove JH, Sant'Anna AC, Mata-Padrino D (2022) Soil health—it's not all biology. Soil Secur 6:100051

Creamer RE, Barel JM, Bongiorno G, Zwetsloot MJ (2022) The life of soils: Integrating the who and how of multifunctionality. Soil Biol Biochem 166:108561

FAO (2011) The state of the world's land and water resources for food and agriculture (SOLAW)—managing systems at risk. Food and Agriculture Organization of the United Nations, Rome and Earthscan, London. Available: http://www.fao.org/nr/solaw/the-book/en/. Accessed 20 June 2022

FAO (2013) Guidelines to control water pollution from agriculture in China: decoupling water pollution from agricultural production. FAO water reports 40. Food and Agriculture Organization of the United Nations (FAO), Rome

FAO and WWC (2015) Towards a water and food secure future—critical perspective of policy-makers. Food and Agriculture Organization of the United Nations, Rome and World Water Council (WWC), Marseille. Available: https://www.fao.org/3/i4560e/i4560e.pdf. Accessed 21 June 2022

FAO (2019a) Soil erosion: the greatest challenge to sustainable soil management. Food and Agriculture Organization of the United Nations (FAO), Rome

FAO (2019b) Outcome document of the global Symposium on soil erosion. Food and Agriculture Organization of the United Nations (FAO), Rome

FAO (2021) Land & water: sustainable land management. Food and Agriculture Organization of the United Nations (FAO), Rome. Available: https://www.fao.org/land-water/land/sustainable-land-management/en/. Accessed 21 June 2022

Ghosh S (2014) The impact of the Damodar valley project on the environmental sustainability of the lower Damodar basin in West Bengal, Eastern India. OIDA Int J Sustain Develop 7(2):47–54. Available: https://papers.ssrn.com/sol3/papers.cfm?abstract_id=2441593. Accessed 21 June 2022

Hassan W, Li Y, Saba T, Jabbi F, Wang B, Cai A, Wu J (2022) Improved and sustainable agroecosystem, food security and environmental resilience through zero tillage with emphasis on soils of temperate and subtropical climate regions: a review. Int Soil Water Conserv Res, in press. https://doi.org/10.1016/j.iswcr.2022.01.005

Huang B, Yuan Z, Zheng M, Liao Y, Nguyen KL, Nguyen TH, Sombatpanit S, Li D (2022) Soil and water conservation techniques in tropical and subtropical asia: a review. Sustain 14:5035. https://doi.org/10.3390/su14095035

Jaishankar M, Tseten T, Anbalagan N, Mathew BB, Beeregowda KN (2014) Toxicity, mechanism and health effects of some heavy metals. Interdiscip Toxicol 7(2):60–72. https://doi.org/10.2478/intox-2014-0009

Karlen DL, Gardner JC, Rosek MJA (1998) Soil quality framework for evaluating the impact of CRP. J Prod Agric 11:56–60

Lal R (2010) Managing soil to address global issues of the twenty-first century. In: Lal R, Stewart BA (eds) Food security and soil quality. CRC Press, New York, pp 6–19

Lehman RM, Cambardella CA, Stott DE, Acosta-Martinez V, Manter DK, Buyer JS, Maul JE, Smith JL, Collins HP, Halvorson JJ, Kremer RJ, Lundgren JG, Ducey TF, Jin VL, Karlen DL (2015) Understanding and enhancing soil biological health: the solution for reversing soil degradation. Sustain 2015(7):988–1027. https://doi.org/10.3390/su7010988

Luino F, Tosatti G, Bonaria V (2014) Dam failures in the 20th century: nearly 1000 avoidable victims in Italy alone. J Environ SciEng A 3(1):19–31

Mateo-Sagasta J, Zadeh SM, Turral H, Burke J (2017) Water pollution from agriculture: a global review—executive summary. CGIAR research program on water, land and ecosystems (WLE). Food and Agriculture Organization of the United Nations (FAO), Rome, and International Water Management Institute (IWMI), Colombo

Mizuta K, Grunwald S, Cropper WP Jr, Bacon AR (2021) Developmental history of soil concepts from a scientific perspective. Appl Sci 11(9):4275. https://doi.org/10.3390/app11094275

Montanarella L (2015) Govern our soils. Nature 287(7580):32–33

Morgan RPC (2005) Soil erosion and conservation, 3rd edn. Blackwell Science Ltd, Malden

NREL (2021) Food security. Natural Resources Ecology Lab (NREL), Fort Collins. Available: https://www.nrel.colostate.edu/research/food-security/. Accessed 21 June 2022

Oldeman LR, Hakkeling RTA, Sombroek WG (1991) World map of the status of human-induced soil degradation: an explanatory note, 2nd revised edn. Global assessment of soils degradation, The Map Sheets. World Soil Information (ISRIC), Wageninen

Oldeman LR, Hakkeling TTA, Sombroek WG (1994) The global extent of soil degradation. In: Green D, Szabolcs J (eds) Soil resilience and sustainable land use. CAB International, Wallingford, pp 99–118

Robinson DA, Panagos P, Borrelli P (2017) Soil natural capital in Europe; a framework for state and change assessment. Sci Rep 7(1):6706

SWC (2022) A history of soil concepts. The profile, weekly email newsletter. Department of Soil, Water, and Climate (SWC), College of Food, Agricultural and Natural Resource Sciences, University of Minnesota, St. Paul. Available: https://swac.umn.edu/news/profile-2022-02-14. Accessed 21 June 2022

Teng H, Liang Z, Chen S, Liu Y, Raphael A, Rossel V, Chappell A, Wu Y, Shi Z (2018) Current and future assessments of soil erosion by water on the Tibetan Plateau based on RUSLE and CMIP5 climate models. Sci Total Environ 635:673–686. https://doi.org/10.1016/j.scitotenv.2018.04.146

USSLS (1968) Richards LA (ed) Diagnosis and improvement of saline and alkali soils. United States salinity laboratory staff (USSLS), agriculture handbook no. 60, USDA. Oxford & IBH Publishing Co., New Delhi

Yang Q, Zhu M, Wang C, Zhang X, Liu B, Wei X, Pang G, Du C, Yang L (2020) Study on a soil erosion sampling survey in the pan-third pole region based on higher-resolution images. Int Soil Water Conserv Res 8:440e451

Printed in the United States
by Baker & Taylor Publisher Services